Oil and Gas Privatisation in Iran

Oil and Gas Privatisation in Iran

Reza Molavi

Durham Middle East Monographs

ITHACA
PRESS

OIL AND GAS PRIVATISATION IN IRAN
An Assessment of the Political Will

Published by
Ithaca Press
8 Southern Court
South Street
Reading
RG1 4QS
UK

www.ithacapress.co.uk
www.twitter.com/Garnetpub
www.facebook.com/Garnetpub
thelevant.wordpress.com

Ithaca Press is an imprint of Garnet Publishing Limited.

First Paperback Edition

ISBN: 978-0-86372-422-0

British Library Cataloguing-in-Publication Data
A catalogue record for this book is available from the British Library

Typeset by Samantha Barden
Jacket design by Garnet Publishing

Printed and bound in Lebanon by International Press:
interpress@int-press.com

CONTENTS

LIST OF TABLES

LIST OF FIGURES

ACKNOWLEDGEMENTS

There are many people without whose support and co-operation this book would not have been possible. First and foremost I would like to thank Professor Anoushiravan Ehteshami of Durham University whose personal interest in my subject, his mentoring and encouragement made this journey a pleasant experience for me. I am also grateful to Christine Bird, President of CHC Helicopters International, and Chris Krajewski, the Vice-President, for their encouragement and support throughout this undertaking.

Much of the fieldwork could not have been carried out without the help of various colleagues in Iran. In particular, I wish to extend my thanks to Seyed Kamal Hosseini and his staff at Iran Economics, for scheduling my interviews within a tight time frame and making sure that my interviewees were available on the dates designated for the interviews. I also wish to thank Ghobad Shirzadi for sourcing articles on privatisation through newspapers and magazines in Iran.

Dr David Kerr deserves special thanks for his help during the earlier stages of writing and for his advice on the political dimension of this topic. His comments on the first three chapters of this work set the tone for my arguments in the final three.

Finally and not least, I owe a debt of gratitude to my family for their support and unwavering encouragement. Their understanding for my frequent absence from home is greatly appreciated; Narges, Babak and Cyrus were all brilliant in coping and putting up with me. Above all, I thank them for creating a genial environment in which the writing of this work could move forward unhindered.

INTRODUCTION

The Purpose of the Study

This study explores the issues surrounding the privatisation of some of Iran's national institutions with particular emphasis on the oil and gas industry. Its aim is to understand some of the reasons for an apparent reluctance within Iran's public and private sectors towards privatisation, and this will be achieved by identifying the factors influencing privatisation policy-making in Iran.

Across the Gulf region privatisation policies are currently being contemplated and in some cases are being carried out, while in Iran a privatisation programme was launched earlier, in 1995, and the concept remains high on Iran's political agenda. However, its origins, the reasons for its popularity, its requirements and management, as well as the social and economic consequences are not that well understood in Iran. Thus, given the current political climate in Iran and the socio-economic crossroads at which the country finds itself, I have embarked on an investigation into the privatisation of its oil and gas industry. An added impetus for this study arose from external changes. From President Bush's statements at the Red Sea Summit in June 2003 and the recent political developments in the Middle East, it is not hard to accept the notion that by 2013 the entire Middle East region might find itself poised to become a 'Free Trade Zone' (CNN, 2003). For this to happen, it has been suggested that the countries of the region will have to embrace market economy concepts and agree to move towards transparency and privatisation, a word often used synonymously with the sale of public assets, deregulation and decentralisation.

Little is understood about the relationship between Iranian politics, society, the economy and privatisation. For instance, while the idea of privatisation is currently popular in the region, and in Iran in particular, the consequences of privatisation are rarely thought through. This is especially true as they relate to standards of services, prices, costs, cross-subsidisation, competitiveness, reallocation of human resources, safety and other components linked to the public interest, not to mention the implications for the private sector and existing government organisations. In particular, the study considers the problems that the Iranian oil and gas sectors face in adopting a privatisation model, and what solutions there are that would fit the unique circumstances in Iran.

The Debate over Privatisation

This study is not taking place in a vacuum. There has been a long-running debate over privatisation in purely theoretical terms in Iran and any analysis of privatisation in the country must take account of the arguments, which encompass a host of realities such as the country's history and culture. While the answer to the question "why consider privatising?" is as old as the subjects of economics and management themselves, there still exist many ambiguities. In support of privatisation Smith wrote:

> In every great monarchy in Europe the sale of the crown lands would produce a very large sum of money, which, if applied to the payment of the public debts, would deliver from mortgage a much greater revenue than any which those lands have ever afforded to the crown . . . [w]hen the crown lands had become private property, they would, in the course of a few years, become well improved and well cultivated (Smith, 1904:348).

This positive slant on private ownership and the view on privatisation in the West has not gone uncontested by Marxist critiques of private ownership. However, in the main, and in many contexts, the basic aims of privatisation are claimed to be the promotion of competition and increased efficiency. In line with these goals, Iran's former Deputy Minister of Economy, Dr Mohammad Khazaee (2003), stated that maximum efficiency would be achieved if the decision-making politicians and public servants were replaced by market mechanisms, and that the philosophy and practice of the private sector should be injected into the public sector.

A strategy to privatise requires the size of the public sector organisation to be reduced and public sector borrowing to be controlled and also reduced so as to make room for unfettered private sector borrowing. Privatisation is commonly believed by its proponents to induce transparency in decision-making. Furthermore, because independent capital markets are likely to be more reliable sources of funding than the politically controlled capital markets, over a period of time privatised companies should be able to raise funds to finance efficiency or expansion in a more flexible manner, and so meet their special needs. In addition, financial markets bring their own scrutiny by analysts and bankers, a discipline not available in government (Pesaran, 2003). Keynesian economists might criticise the argument on the grounds that the assets of public enterprises are almost invariably undervalued because of government fears that sales might fall through. Potential conflict of interest with the appointed consultants in the privatisation programme is cited as another difficult aspect (Wiltshire, 1987). The main criticisms of the British privatisation programme, for instance, have been in five major spheres (Wiltshire, 1987):

1) The failure of the government to ensure competition in industries where privatisation occurs, thereby appearing to convert public monopolies[1] into private monopolies;
2) the creation of inadequate regulatory frameworks in such monopolistic or oligopolistic industries, especially in relation to the fulfilment of social obligations on the part of privatised industries;
3) failure on the part of the government to adequately safeguard the national interest against foreign influences in the privatisation process;
4) the revenue from the sale of public assets being apportioned to consolidated revenue rather than being earmarked for future generations and future public uses;
5) the standard of services after sale or contracting out.

In light of the British experience (and others'), the question arises as to whether the government of Iran will find the courage to give up its control over the National Iranian Oil Company (NIOC). Will it be able to successfully reallocate surplus human resources when the NIOC is downsized after privatisation? More will be said about the Iranian case later.

Differences in behaviour under state and private ownership may be compounded by differences in objectives. In the private sector, although profit may not always be rigorously and consistently pursued, there is, nevertheless, a clear bottom line. If losses continue to be made, the firm will ultimately fail. In the state sector, however, objectives are vague and tend to change according to the political climate, leading to uncertainty about long-term strategies within industries (Martin and Parker, 1997). It is also widely believed that public ownership reduces the incentive to secure profits and removes the threat of bankruptcy. Private ownership would undoubtedly provide a significant spur to managerial efficiency. Capital would be allocated according to the market, rather than politics, and interventions would largely cease. To secure such advantages, Enoch Powell, the right-wing British politician, and Milton Friedman, the famous monetarist, advocated immediate and wholesale denationalisation (Beesley and Littlechild, 1989).

Background to the Privatisation Policy in Iran

To analyse the relevance of the privatisation of the National Iranian Oil Company and to understand the idiosyncrasies and hesitation of the Iranian policy-makers, one must look into its immediate background. Most oil-rich Third World economies have had difficulty in evolving into true liberal societies. By owning or controlling revenues generated by oil, the state is able to dominate society, making all classes and groups economically dependent on their "black gold". Oil tends to centralise state power in the Middle Eastern countries as this commodity forms anywhere between 80% to 93% of their foreign exchange earnings (USEIA, 1998, 2005). When the US-backed Pahlavi regime began to fall apart, problems within Iranian society were compounded by the diplomatic pressures applied by the US, its main ally. To make matters worse, the decline of oil prices in 1974 sharply curtailed its economic ambitions (Kamrava, 1990). A quarter of a century after the fall of the Shah of Iran and the establishment of an Islamic regime, the country began to re-examine its revolution. The 1979 revolution was originally grounded on a fight for democracy, civil rights and freedom from foreign interference. But the revolutionary process was soon dominated by Ayatollah Khomeini and a coterie of clerical activists, who imposed, *inter alia*, strict Islamic dress on women and banned alcohol and music. The seizure of the American Embassy and the extremely brutal repression of opponents turned Iran into a pariah state.

President Mohammad Khatami tried to restore normality, both in Iran and in its relationships with foreign countries. Elected with 69% of the vote in the May 1997 elections, and again, with some 60% of the vote in the 2001 elections, he certainly enjoyed the mandate of Iran's youthful population. More than 40% of the country's 65 million citizens were under the age of 15. However, the job market was tight and, with the introduction of the United Nations' second set of sanctions, the importation of desperately needed spares for the ageing oil and gas industry became much harder. Having the second largest bank in Iran, Bank Sepah, on the sanctions list, compounded the problems of transacting with the sources of supply in US dollars. Even if the tightening of the noose around the financial institutions did not make it impossible to secure the requirements of the country from abroad, it certainly made it uneconomic to buy goods and services from the traditional sources of supply.

The economy was in serious trouble. Iran's population had doubled since the revolution, dramatically increasing consumption. With no growth in the output of the economy (GDP), the standard of living had halved. Half the population, including 70% of government employees, lived below the official poverty line and still do. Many have more than one job just to make ends meet. Unemployment was estimated at 25% and the future seemed bleak. Every year, 800,000 young people entered the job market. It is estimated that more than $100 billion were needed to be invested to maintain existing employment levels, but only $2.3 billion were allocated to job creation (Reuters, 1998a).

In 1998, fuel subsidies alone reached $11 billion per year – double the country's development budget. Food subsidies on basic commodities such as bread, rice and sugar consumed another $2.2 billion per year. Together, these subsidies used up virtually all of the country's $16 billion oil revenue for the year (Reuters, 1998b). President Khatami accepted the need for market reform and, with no other source of capital available to the country, he sought to attract foreign investments. The President skilfully combined cuts with subsidies, the privatisation of state firms and changes to labour laws; in short, he applied shock therapy to the market and called for the liberalisation of social and political rules to allow freedom of expression, freedom of the press and the establishment of a civil society. At the same time he became fully committed to restructuring the economy. His government approved the privatisation of 538 state-owned companies through auctions and stock sales, with plans to privatise another 2,000 (Reuters, 1999). The Iranian Parliament (Majlis) approved plans to overhaul the oil industry and boost Iran's sagging oil production from 3.5 million bpd (barrels per day) to 6 million bpd over the next ten years (USEIA, 1998). This required at least $120 billion in capital, which had to come from private investors and foreign sources.

Through a strategic development plan and the implementation of various petrochemical projects, efforts were made to raise the production level of Iran's petrochemical industry. Begun in 1997, the plan was divided into five phases, to last until 2013. By the end of this period the total volume of final products was estimated to reach 16.8 million metric tons per year. The value of investment amounted to $20.6 billion, with total sales topping $11.8 billion (PIIC, 2000).

Participation by the private sector, whether in the form of investment in stock or in the form of launching new projects, can have a considerable impact, and it was for this purpose that the Petrochemical Industries Investment Company (PIIC) was established. As privatisation became one of the main policies of the government, it led to the restructuring of the National Petrochemical Company (NPC): plants with a capacity of less than 100,000 tons per annum were subject to transfer to the private sector (PIIC, 2000). The private sector was also allowed to invest in medium-sized petrochemical plants. NPC offered bonds for the construction of petrochemical plants, and also supplied raw materials to the private sector at a 30% discount compared with world prices.

Providing macro-economic stability and ensuring progress in the key area of the economy was critical in putting Iran on a sustainable growth path, announced Tahmaseb Mazaheri, the former Minister of Economy, after a meeting with his UK counterpart, Gordon Brown, at the end of June 2003. To create a suitable environment for the private sector to function properly, coupled with reform of the financial sector and pricing system, Iran must rid itself of the notion that foreign investors are imperialistic. Although the Ministry of Energy authorised the sale of 14% of the capacity of its state-owned power plants (Utilipoint, 2004), NIOC remained off-limits. This study investigates whether Iran has any choice but to privatise NIOC while taking into account reasons for the reluctance on the part of the people and the government.

Since the announcement of the privatisation programme in the Five-Year Development Plan (FYDP) in 2000 (Privatisation Organisation, 2002), significant progress has been made and a new agency set up by the Ministry of Finance. The Majlis has promulgated regulations governing privatisation. The Privatisation Committee (PC), under the supervision of the president, reviewed 1,039 public sector enterprises for privatisation: 217 were to remain public, 87 were to be liquidated and 735 were earmarked for privatisation. During 2000–01, the government sold Rls2,040 billion (Rials) (about 0.5% of GDP) worth of shares in public enterprises on the Tehran Stock Exchange (TSE). It ceded Rls100 billion worth of shares to workers. Rls1,800 billion worth of shares were put up for sale. In the oil sector, the Oil Ministry announced that the PC would cede 23 subsidiary firms to the private sector over the FYDP period, including the National Iranian Tanker Company (NITC) and the Liquefied Gas Distribution Centres (LGDC) (World Bank, 2001).

The critical issue facing Iran today is its stand-off with the international community over its development of nuclear technology. While this point is self-evident, what is not so obvious is that the eventual resolution of this crisis is likely to profoundly affect Iran in the future, not just in terms of its economy and its privatisation policies, but also in terms of its foreign policy, domestic politics and, potentially, even the nature of the state itself.

Iran's economy was in desperate need of radical reform. Mahmoud Ahmadinejad came to power, in June 2005, on the back of an anti-corruption campaign and the promise to reform the economy, proclaiming that he would bring the oil money to every table in Iran. In November 2005, he announced that the government would begin to distribute shares of Iranian publicly held industries to the entire populace. Rumours quickly spread that this was a preliminary step to the seizure and redistribution of many private industries and other assets, and the result was a vast, sudden flight of capital out of the country, possibly in the order of $200 billion. This, together with his remark about wiping Israel off the map, caused the Tehran stock market to collapse.

Almost 40% of the Iranian GDP is accounted for by the Bonyads, which are nominally charitable foundations established to administer the Pahlavi regime's assets on behalf of the Iranian people, but in actuality are massive corruption machines that bankroll the senior leadership. Corruption reduced liquidity, frightened off investment, boosted inflation, spurred widespread unemployment, diminished non-oil exports, impoverished the middle class and created a very serious gap between the rich and the poor. Iran's latest Five-Year Plan (Privatisation Organisation, 2004) called for $20 billion investment each and every year in addition to $70 billion to recapitalise Iran's decrepit oil industry. With corruption so rampant it was difficult to see where the money would come from.

According to Amini (2004), and expounded further by Nili *et al.* (2004), Iran's problem is that there are only three capital markets in the world capable of generating such levels of investment over the next five to ten years – these being the United States, Europe and Japan. Much as Iranian hardliners would like to believe that Russia, China and India could replace the West, the fact of the matter is that they will not be in a position to do so for about a decade. In addition, there is the issue of superior Western technology: the Iranians would much prefer to have Exxon or Shell repairing the oil infrastructure, rather than Lukoil or Sinopec (Pollack, 2006).

Iran has recently issued an executive order for the privatisation of 80% of several state-owned companies, but retained firm control over the upstream oil sector and key banks. Ayatollah Seyed Ali Khamenei, Supreme Leader of the Islamic Republic of Iran, said in his order: "Ceding 80 per cent of the shares of large companies will serve to bring about economic development, social justice and the elimination of poverty" (Roshanzamir, 2006).

The executive order was designed to revive Iran's stalled privatisation programme and kick-start the country's many uncompetitive industries of the Bonyad Mostazafan, which are heavily protected and are the root cause of ill-feeling among the various sectors of society.

The implementation of this action plan would change the government's role from direct involvement in ownership and management of large companies to supervising and guiding different sectors of the economy in line with the regulations of the World Trade Organisation. Ayatollah Khamenei said that the privatisation process would help reinforce the private sector in the national economy and encourage companies to compete in international markets and that the downstream oil and gas sectors would be privatised, but excluded the upstream oil and gas industry, NIOC, the state companies involved in exploration and the production of crude oil and gas. The former head of Tehran's Chamber of Commerce, Mohamad Reza Behzadian, noted that the move amounted to a major change in the old constitutional attitude toward the economy as the absolute property of the government, stating that this order means a change in the government's status from owner to guide (AFP, 2006).

In early March 2007, the bitter rivalry between Iran's President, Mahmoud Ahmadinejad, and the country's leading elder statesman,

Ayatollah Hashemi Rafsanjani, erupted into a public struggle for control over economic policies. Hashemi Rafsanjani, the President's most influential opponent, set the scene for a power struggle by telling Iranian journalists on 7 March 2007 that Mr Ahmadinejad's "trial period is over". He said he would use his position as head of the Expediency Council (a state body empowered to set the Islamic regime's long-term goals) to reshape the government's economic policies.[2] This is an extract of what he said to journalists:

> Now the trial period is over and the supervising role of the expediency council should be enacted more seriously. Under the 20-year outlook plan the country's reliance on oil should be reduced by more than 10% each year but during the last two years this process has been reversed. Next year's budget depends on oil to an even greater extent than those of the last two years.

It is rather obvious that while there was unanimity and abundance of political will in carrying out the plans for economic reforms, particularly in respect of the privatisation programme, there were no signs of universal acceptance of how the political reforms would be made to keep step with the economic liberalisation plans.

Determinants of the Privatisation Policy

The previous section has surveyed the background and history of privatisation policy in Iran and the aims and objectives of the study. As already intimated, no published research on the privatisation of the oil and gas industry has been conducted outside Iran. This study aims to fill this gap and provide future researchers with the empirical data collected by the author during personal interviews and by reviewing existing archives in Iran. We are now in a position to outline the question which this study specifically addresses. It aims to identify the determinants of the Iranian policy of privatisation in order to answer the question of whether there is the political will to privatise the Iranian oil and gas industry. But to answer this question, it is necessary to address certain sub-issues:

- Privatisation in theory and practice:
  - Contemporary privatisation theory;

    - Other similar attempts at privatisation, i.e. privatisation in comparable political, economic and social environments.
- How the historical evolution of Iranian politics and economics affected the current policymaking environment.
- What the structural features of importance are, including previous privatisation transactions; the legal framework; the political, social and economic environment; and regional and global geopolitical and economic factors.

Furthermore, certain facts about Iran's politics and society also have to be addressed. These include:

- The Iranian domestic environment:
    - Past policy evolution;
        - domestic policy
        - economic policy
        - oil policy.
    - The social psychology of Iranian government and society;
        - the role of revolutionary ideology
        - the role of religion.
- The international environment, including:
    - Relations with Middle Eastern states;
    - relations with the West;
    - relations with international organisations.

These issues are all addressed in this study. The following section summarises the chapters and the methodology used in this research. As regards the latter a three-pronged approach is adopted. First, a historical survey of Iran since World War II is provided in order to build the political and economic context for the determinants of privatisation policy in Iran. Second, a theoretical basis for the determination of privatisation policy is developed and the precedents of privatisation policy are explored. Finally, the specific background and the legal and institutional framework as well as the policymakers' perspectives are described and analysed. Together, these three approaches may provide an understanding of the determinants of the privatisation policy in Iran.

Summary of Chapters

Chapter 1 presents an overview of the theoretical basis behind privatisation in general. It describes the various schools of thought concerned with privatisation and sets out the realm of possibilities that privatisation can be conceived for Iran. To illustrate this realm of possibilities, the chapter then looks at some concrete examples of privatisation in countries surrounding Iran, including Pakistan, Azerbaijan, Kyrgyzstan and India. The experiences of two other countries further afield, namely Egypt and China, are also examined since they have a major effect on the Iranian economy. In other words, the chapter examines the possible privatisation decisions which can be made by the Iranian policymakers. This is the case because it must be assumed that as rational players, Iranian policymakers, while interacting with other policymakers outside Iran, can only make decisions based on historical or comparative precedents. This sets the stage for examining the types of privatisation to adopt, and in that sense it might be considered as providing one of the determinants of privatisation.

Chapters 2, 3 and 4 employ a political economy approach to analyse the history leading up to present discussions of Iranian oil and gas privatisation. Each chapter explores an era of Iran, beginning in Chapter 2 with the postwar period, dominated by the Shah, and leading up to the 1979 revolution. Chapter 3 then addresses the period roughly equal to Khomeini's rule, from 1979 to 1989. Chapter 4 addresses the contemporary Iranian situation, from 1989 to the present. Each of these chapters explores political issues, including domestic and international politics. They also review economic issues, including changes in the economic outlook of each era. The chapters then analyse the interplay between these factors.

Throughout this study, the political economy methodology is utilised. It admits a causal relationship between economic change by political change and vice versa. It is here assumed also that domestic factors have international consequences just as international factors have domestic consequences. In the case of Iran, we have seen that as the price of oil rises, due to international events and demand, the privatisation fever subsides, but as prices drop the attraction of privatisation increases among the political elite.

Clearly, the interplay between domestic and international political and economic factors have decisively influenced Iranian privatisation policy. Each of these (four) factors is summarised as follows. The first is the disjunction between Iran's government and society, one that

Katouzian (2000; 2003) has called the "theory of arbitrary rule". It existed in pre-revolutionary Iran and continued thereafter, though in a different form. Such disjunction between the state and society is the first determinant of policy, and privatisation policy more specifically. The second factor is the disjunction between economic and socio-political modernisation. This is examined in relation to the Shah's attempts at economic and political reform and similarly in post-revolution Iran. The third factor determining policy is Iran's dependence on oil. Oil has played a major role in the life of Iran under the Shah and has continued to do so since the revolution. It has distorted as well as given bias to both domestic and international policy of the state and still plays a huge role in determining privatisation policy. The fourth factor to consider is the relations of Iran with the West (the US in particular) and Iranian society's perception of these relations. In particular it perceives the West's attitude towards Iran as imperialistic, and this had a significant effect on the nature of Iran's policy before the revolution and since.

These four factors have all helped to determine the nature of Iranian policy with regard to privatisation throughout the history of Iran, going back at least to the days of the Shah. They continued into post-revolutionary Iran due, in part, to the brevity of the Iranian revolution. In addition to these determinants, there are two supplementary factors which have had a significant influence since the revolution. In some senses these had existed before the revolution, but became much more pronounced after it.

One of these was the role of the Islamic Republic's revolutionary ideology. It led to changes in the structure of government and in foreign policy. In respect of the latter one may cite the hostage crisis and Iran's handling of the Iran–Iraq War and the Gulf War of 1990. Another key issue has been Iran's development of its nuclear capability. Together these factors have, broadly speaking, contributed to changes in government policy and, more specifically, to privatisation. The other factor was the role of religion. In particular, religion has had a complex, but direct impact on privatisation policy (this aspect is explored in some detail in the course of this study). Needless to say, religion had other effects on politics and economics; one may also mention the dichotomy between Khomeini's views on the revolution on the one hand and his views on fundamentalist Islam on the other. This has had indirect, though no less wide-ranging, effects on Iran's privatisation policy.

Chapter 5 focuses on the four primary factors which directly affect privatisation policy. These serve as background to a number of determinants: the legal and institutional framework; the nature of oil and gas relative to privatisation; and the thinking of the relevant policymakers. The chapter pulls together all of the historical and theoretical contexts and determinants of Iran's privatisation policy.

The narrative on the legal and institutional framework is based on a close reading of the relevant legal documents concerning privatisation post-1990 in Iran. It also explores the institutional issues which have arisen since that time and which constrain policy decisions. The chapter brings together the theoretical arguments regarding privatisation (discussed in Chapter 1) and the legal and political evolution discussed in Chapters 2, 3 and 4. The third factor – the oil and gas industry – is a unique feature in Iran, and its privatisation needs to be addressed separately from other types. It also brings into play supplementary information to help better understand the characteristics of this sector. With respect to the personal views of decisionmakers concerned with privatisation in Iran, these were obtained through extensive interviews. The questions put to the interviewees were thoroughly prepared beforehand so as to provide a perspective on what kind of information was relevant and what sort of bias can be expected. The findings are then described and examined.

Chapter 6 summarises the conclusions of each chapter and brings out the implications of the study. The factors determining the privatisation policy, particularly those concerned with the oil and gas sector, are illustrated by means of Table 6.1. They all interact to determine the privatisation policy in the country, with each playing its own role and so must be discussed separately. As seen in Chapter 5, the contextual elements contribute to the specific factors which determine privatisation. These conclusions enable us to make some hypotheses concerning, *inter alia*, the prospects for the privatisation of the oil and gas in Iran. The framework developed in this study allows us to do more than simply make guesses about such prospects.

A Note on Source Material

This study uses a dual methodology based on primary and secondary sources. As illustrated in the summary of chapters above, secondary sources are combined with publicly available primary sources so as to provide a context for answering the question raised in this study. In addition, fresh and significant information regarding the personal views of policymakers (who are associated with the privatisation process) has been extracted, and is set out in Chapter 5. The primary and secondary sources include:

- relevant legal texts;
- government documentation;
- regulatory documentation;
- banking and other business documentation and data;
- speeches by government officials;
- theoretical and other academic works and articles in journals.

Face-to-face interviews with key figures involved in the privatisation process gave a subtle picture of their views and attitudes regarding privatisation in Iran. These interviews, backed by a questionnaire, provided a nuance to the study without which the human element would have been submerged. My encounters were mainly with key personnel of the Ministry of Power and affiliated companies within the oil and gas sector, including the National Iranian Oil Company (NIOC), the National Petro-Chemical Company (NPC), the National Iranian Tanker Company (NITC) and the Helicopter Services Organisation (HSO). (For a list of the persons interviewed, see Appendix 3.)

The questionnaire is designed with national objectives in mind and forms part of my method of collecting data. Case studies on the subject from other countries were brought to the attention of interviewees during the discussion of the questionnaire. Collection of unpublished data can best be made through personal visits and interviews. Having worked as a consultant and provider of services to the oil and gas sector, I have had access to decisionmakers for personal interviews. In this connection, it is worth mentioning that the replies to my questions during these interviews were not taken at face value, but were sifted out from the interviewees' personalities.

NOTES

1 One might also refer to them as 'regulated monopolies', such as the utility companies in the United States.
2 Robert Tait in Tehran, Wednesday 7 March 2007, *The Guardian*.

1

PRIVATISATION IN THEORY AND PRACTICE

Introduction

Rather than probing into the background of decisionmaking in Iran, this chapter will look at the precedents in other countries ranged before Iranian decisionmakers. Indeed, the theoretical and precedent bases for possible decisions are not determined by any individual player, but rather by a structural feature of the international system, the interplay between different players, and, crucially, by the evolution of ideas, what Richard Dawkins (1991) calls memes. Thus it is the realm of possibilities which inspires as well as constrains policymakers when taking and implementing decisions. It includes the theoretical arguments which practitioners could use as a framework for privatisation in Iran. Equally important are the precedents provided by other states. These were also constrained by theories about privatisation existing at the time as well by the experience of others, along with their individual circumstances and the resources available to them.

There is an interplay between the possible decisions which Iran can make concerning privatisation and the actual decisions taken over time. This is a process called structuration, developed by Giddens (1984), to serve as a means of solving the so-called "agent–structure" problem. The problem is that it is not clear whether structures or systems cause agents to act as they do, or if agents create the structure. Giddens' solution is to say that there is an interplay over time between the structure and agent where each contributes to the formation of the other. In the case of Iranian privatisation, the structuration solution is that theory and precedents (the structure) are informing the decisions of policymakers (the agents), and that at the same time their decisions are contributing to the nature of the structure, i.e. by becoming another precedent. The implication of this for Iran is that the theory and precedents of privatisation must be studied carefully.

This chapter begins by addressing the theory of privatisation, providing in the first instance a very general overview of the issue, including definitions of important concepts. It then moves on to illustrate and examine the different possible forms of privatisation. This sets the stage for a more in-depth examination of some specific issues that arise in the privatisation process and which have important implications in Iran; in particular, the issue of the right to property is then explored.

Secondly, the chapter surveys the experiences of the (relevant) states surrounding Iran and those further afield which have considerable impact on Iranian privatisation. The countries surrounding Iran include Pakistan, Azerbaijan, Kyrgyzstan and India. The non-neighbouring states, which have a major effect on the Iranian economy, are Egypt and China. After casting a glance at the history of their privatisation attempts over the last decades, the discussion moves on to analyse these attempts, specifically in regard to the structural constraints they impose on other states in their own attempts to privatise.

Privatisation and Related Concepts

The term privatisation has a wide range of meanings which can be seen from the large number of theoretical and empirical analyses on the subject (Demsetz, 1967; Backhaus, 1989; Barzel, 1989; Bouin, 1992). One particular study, for example, has enumerated as many as 15 understandings of the concept of privatisation (Gormley, 1991). In more specific and fundamental terms, it is understood as the transfer of property and controlling rights of enterprises, infrastructures and other economic establishments from the state to the private sector. In a wider sense, privatisation is conceptualised as the process of enhancing the effectiveness of the non-state economic sector. Obviously, the latter can both incorporate and exclude the former, and as such could open up possibilities for rhetorical commitment to privatisation on the part of the political and economic elites.

In an even wider sense, privatisation can be interpreted as a process which incorporates broad measures of economic liberalisation, such as reducing state paternalism, the removal of bureaucratic obstacles for economic agents and the improvement of the functioning of state-owned enterprises. In this sense, even if privatisation is being implemented in the economic milieu, it represents a *bona fide* political process which requires the researcher to delve deeply into its political determinants and consequences. As such, privatisation clearly differs from other phenomena signifying a withdrawal of state influence and control over the economy, such as commercialisation, decentralisation and deregulation. They can all exist independently, but can also join into a coherent governmental privatisation strategy.

Commercialisation implies the extraction of state-owned enterprises from the framework of state-planning, i.e. they become independent economic agents with their own capital and juridical standing that function according to the logic of the market. Decentralisation is defined as the process of transferring ownership and controlling rights from central to local structures of state authority, whereby state planning becomes a mere summary of local economic planning. Finally, deregulation implies the withdrawal of the state from many aspects of the country's economic life through, *inter alia*, the liberalisation of prices, the liberalisation of the trade regime, demonopolisation, etc.

The notion of privatisation should also be distinguished from the closely related, but narrower, concept of divestiture. It presupposes the spin-off of filials, branches or units of a given enterprise, either because of hardship or economic necessity. In the first case, the revenues from the sale would help maintain the rest of the activities; in the second, there would be a restructuring or re-orientation predominantly towards other areas of the business. Divestiture can be effected through the transfer of shares or material assets and is often a method of exiting a position of market monopoly, voluntarily or not. Another associated mechanism, which also falls short of the transfer of property and controlling rights, is that of contracting service providers, such as managerial or consultancy services, and outsourcing certain of its less profitable activities to third parties (Gormley, 1991).

Forms of privatisation

The process of privatisation, broadly defined above as the transfer of property and controlling rights over state-owned enterprises, can take multiple forms. They are all part of this definition and operationalise it from different perspectives and according to different criteria, including scope and method, parties involved, type, size and sector of the privatised enterprise, market share, etc. Table 1.1 provides a concise presentation of the various forms, which are further elaborated below.

The different forms shown in Table 1.1 generate a different focus on the determinants and consequences of privatisation. Each implies a specific configuration of relevant actors, interests and institutional frameworks, which define a particular approach towards its analysis. In terms of its scope, privatisation can be full or partial. The former implies the complete withdrawal of the state from the privatised enterprise, while the latter presupposes the preservation of a certain level of state ownership (usually a minority stake). The latter option is preferred when the industry within which the enterprise operates is placed under sustained governmental control, as the new enterprise owners are able to ensure a continued interest on the part of the state in its progress. This comes with the price of having to take into account certain economic and social policies that the government undertakes to fulfil.

TABLE 1.1
Forms of Privatisation

Criterion of Distinction	**Form of Privatisation**			
Scope of privatisation	Full	Partial		
Privatisation method	Public offering of shares	Tender	Negotiations with an investor	
Investor	Foreign	Domestic	Combined	
Seller	Specialised state organ	Governmental ministry	Intermediary	The Enterprise Board
Type of state ownership	Restructured enterprise (ownership of shares)	Non-restructured enterprise (no division in shares)		
Owner	Central government	Local authorities		
Volume of the privatised assets	Large	Medium	Small	
Economic sector	Industrial privatisation	Bank privatisation	Agriculture, services and trade	
Market share of the enterprise	Monopoly	Non-monopoly		

Source: Based on Pirie (1988) and Vuylsteke (1988).

There are generally three ways of transferring state ownership into private hands. These are: the offering of shares to the public so that ownership is dispersed as widely as possible, or by tender or by negotiation, both of which suppose a single buyer. The method pertaining to negotiation offers most flexibility in terms of being able to field additional proposals following the initial presentation of the standpoints of both the investor and the seller. In terms of investor orientation, privatisation can be aimed at attracting foreign buyers or at stimulating the acquisition of economic assets by domestic companies. Depending on the specific privatisation policy, a combination of these two types of investors in a single joint venture company can also be acceptable to the seller.

With regard to the seller, privatisation can be implemented by a specialised state organisation, a governmental ministry, an intermediary or, in Iran, by the Enterprise Board. Among these (in cases of ownership by the central government, and not municipalities), apart from the ministry, there is no overlap between the seller and the new owner and this may allow further dimensions in the analysis of any particular contract. In terms of the type of ownership, broadly, most companies can be owned through their shares, where the state possesses the shares of a company whose structure requires their existence; or it can be owned as a material object such as production assets and buildings. Both types imply different privatisation strategies.

The rest of the forms of privatisation are more or less self-explanatory, distinguishing between ownership by the central government and municipal authorities, among large, medium and small privatisations (in cases of large-scale privatisation across the country), among different economic sectors, and between a monopoly and non-monopoly market share of the privatised enterprise. An interesting form of privatisation, where the state had acquired enterprises through nationalisation without due compensation, is the so-called restitution. If undertaken at all, a policy of restitution has the potential to permeate all of the above forms of privatisation and generate policy proposals that need to be accounted for in a comprehensive way (Ohashi, 1980).

Right to property

The right to property, its conceptualisation and implementation within any given society, is fundamental to understanding the potential content and orientation of the policy of privatisation pursued by its political and economic elite. It relates to the fundamental dilemma facing policymakers when privatisation is being considered; there seems to be an inherent conflict between a desire to improve efficiency, on the one hand, and a disposition to maintain continuity of the economic and social policies they have been pursuing in the wider interest of the community, on the other (Mihaly, 1992). Thus the different dimensions of the right to property in privatisation policy, contribute to both exacerbating and alleviating this conflict; the following paragraphs demonstrate how.

The basis of property rights was laid down by Roman law. It consists of (a) the right to use and manage (*usus*); (b) the right to revenue (*usus fructus*) and (c) the right to sell or exchange, i.e. fully or partially release oneself from property rights against a certain price (*abusus*). In addition to these, a fourth element has also been seen as relevant, particularly with regard to privatisation, and this is the right of residual control of the property. This right is different to the right to use and manage and the right to sell and exchange, and implies control over all decisions regarding the use of the property which are not explicitly regulated by law or contract (Munzen, 1990).

This dimension of property rights offers a key analytical space for a new way of determining which agent (economic or political) is the owner of a given enterprise. A creditor to a company acquires short-term rights to revenue. Since this forms part of property rights, the argument could be advanced that the creditor is thus an owner as well. Indeed, privatisation through the purchase of company loans has been utilised by interested investors. However, considering the dimension of residual control, a creditor has no comparable standing to that of the proper owner. In any given political and social context, this right is regulated by the constitution and (where applicable) decrees by the country's rulers. In their totality they form the property regime in that particular setting and are an important focus of study and analysis.

The various forms of privatisation cover all four elements of the right to property, either together or separately. An example of the latter is a contract for management with a purchasing clause (a concept close to that of financial leasing), which implies an initial transferral of the right to use and manage and the right to revenue with a subsequent transferral of the right to control and (possibly) sell or exchange. It is conceivable that privatisation policies only envisage a partial transfer of property rights or impose certain limitations on the new owner, particularly when this is seen as more politically acceptable than full and unrestricted transferral by those taking the decision. Table 1.2 summarises some of the common limitations employed when privatising state-owned enterprises.

TABLE 1.2
Limitation of Property Rights during Privatisation

Rights	Limitations		
Right to use and manage	a) obligations to preserve the area of activity	b) limitations on relations with third parties	c) limitations on the percentage of owned shares
Right to revenue	a) obligations for committing to an agreed investment programme	b) obligations for maintaining agreed levels of employment	c) obligations to pay off existing debts
Right to sell or exchange	a) limitation on the volume of assets that can be sold on to third parties	b) limitation on the volume of shares that can be sold on to third parties	

Source: Vickers and Yarrow (1989).

Most often, the above limitations are the result of a professed governmental policy regarding the future of the privatised enterprise. Usually, it is the privatisation of enterprises of great significance for the economy of the country that is subject to limitations. There could also be cases where limitations are imposed for ulterior motives, for example, to favour a certain buyer, or to frustrate attempts to privatise a certain enterprise. In the analysis of the imposition of restrictions on the transfer of property rights, it is often hard to distinguish rhetoric from interest. While taking account of this difficulty, the following paragraphs outline the more obvious functions that each of the above limitations is meant to serve.

The obligation to preserve the area of activity is prompted either by the lack of sufficient number of similar businesses or to prevent the diversion of assets which were created for a particular purpose to other activities. The limitation regarding the relations with third parties is aimed at limiting the access of certain companies or individuals into the particular sector of the economy where privatisation occurs. The limitation on the percentage shares owned by a single investor is designed to avoid a high concentration of ownership within the domestic economy and hence prevent monopolies from forming.

The obligation to commit to an agreed investment programme is often prompted by the development strategies of the government aiming to boost a certain sector of the economy, or a certain enterprise that is technologically lagging behind its competitors, or one that is important to the overall economy of the country. The requirement to preserve/enhance the employment level is usually imposed on investors in enterprises that are either major employers or operate in regions where employment is an important element of the governmental social policy. The obligation to pay off the debts of the privatised company is imposed on those enterprises that have accumulated large debts, particularly if those are due to state-owned agencies which the government is not prepared to transfer to the state budget.

The history of ownership goes back a long way. Human beings have aspired to increase and expand the spectrum of their dominance by extending and enhancing their domain of proprietorship. As compared to the secular systems, Islam has varying philosophies and views about the extent and nature of the right of individuals to ownership. The Qur'an states:

> You who believe, do not use up one another's wealth to no good purpose, unless it is for some business based on mutual consent among you. Do not kill one another: God has been Merciful towards you (4:29).

Another verse says:

> You who believe, do not enter any houses except your own homes unless you are sure of their occupant's consent, and greet them politely (24:27).

Rafi' b. Khadij reported that the Messenger of Allah (SAAS) said:

> One who cultivates on the land of other people without their permission, has no right on the yield except to the extent of his expenses (Siddiqi, 1988).

TABLE 1.3
Definitions of Basic Islamic Concepts

Notion	Islamic Concept
The relationships of human beings with God	• Everything belongs to God, even human beings themselves. • Man is the vicegerent of God on this earth. • The key to Islamic philosophy lies in man's relationship with God. • Man holds property in trust for which he is accountable to God. • Different aspects of the Islamic social framework are united by one principle of *tawhid* (the oneness of God). • With regard to the relationship of man and society to God, emphasis is laid on both purification and growth (*tazkiyah*).
Pursuit of wealth	• In Islamic philosophy, there is no limit to the earning of wealth within the boundaries of Islamic principles.
Class structure	• Emphasis on fair distribution of wealth and welfare.
Social justice	• Emphasis on fair distribution of wealth. • Emphasis on human welfare and security.
Spiritual uplift	• Emphasis on moral values. • The fulfilment of spiritual and material needs of individuals and society necessitates a vital role of the state. • The spiritual and material goals of a Muslim society can only be achieved if all the people in the society utilise their God-given talent with full efficiency and honesty.
Concept of development	• Development includes moral, spiritual and material aspects.
Ownership	• Absolute ownership by man is a concept alien to Islam as it belongs to God. There are definite obligations towards others attendant upon the individual rights of ownership. The Iranian Constitution, which claims to be based on *sharia*, states that there are no prohibitions against wealth as long as it is legitimately acquired (Article 49).

Source: Ahmad (1980), Siddiqi (1988), Naqvi (1989), Mannan (1989), Ahmad (1991), Ahsan (1991), Chapra (1992), Ahmed (1996), Rima (1996), Choudhury (1998).

Whilst the Qur'an makes it explicitly clear that the ultimate owner is God and only God, the right of the individual to hold property is accepted and respected (see Table 1.3). The dichotomy of this God–man relationship in terms of ownership is reconciled by the notion of trusteeship of man, and man's accountability to Him. Therefore, a state representing the will of God – an Islamic state such as Iran – may impose limits on individual property rights. According to the distinguished Iranian jurist Ayatollah Taleghani, the Islamic economy "bears distinguishing features, as it generally recognizes individuals as owners of the fruits of their labour and free agents in transactions, within the framework of the relevant ordinances" (Taleghani, 1983:34).

Beyond these limits, it is the ruler (Islamic State) which functions as the owner or supervisor in regard to the public capital and exchange regulations within the public sphere. Referring to an Islamic-oriented economy, he writes:

> [It] is not based on an unlimited freedom to ownership, nor is it based on the public ownership that results in the total denial of individual ownership and freedoms . . . Property ties and economic relationships are bound with human sentiments and instincts, and the social milieu reflects the totality of these ties. Islam envisions distribution, like production, as the natural and innate right of the one whose labour is involved, such that the individual is free to choose his labour, and labour is the source of the right of ownership (Taleghani, 1983:34).

In Table 1.4 the history of Western economic thinking is summarised here for two main reasons. Firstly, it can be seen that there is no consensus of view. Instead, there is a series of complex and continuing debates. Although it is sometimes fashionable to agree that there is a telos, a tendency for a single meaning to emerge in these debates, this is not really the case. Nor is it the case in Islamic economic argument. Secondly, as this study shows, ideas about privatisation have interacted. It is perhaps going too far to say that there is a "conversation" between Western and Islamic countries regarding privatisation, but there has been at least some dialogue. This book tries to promote a serious exchange between Western and Islamic thinking.

TABLE 1.4

Overview of Wealth and Ownership

School of Thought	Description
Pre-classical **Plato &** **Aristotle**	• They believed that a minimum amount of wealth is essential to the good life. The household exists for the purpose of satisfying natural wants by producing useful commodities or acquiring them by exchange for consumption. • **Plato's view:** 1. A systematic expansion of the principles of society and the origin of the *city state* and the division of labour; 2. The city arises because of the division of labour, which itself is the result of natural inequalities in human skill and the multiplicity of human wants. • **Aristotle's view:** 1. More private property with its liberal uses. 2. In favour of money rather than barter or exchange. 3. Against usury. 4. His attack on communal property is almost entirely based on the *incentive* argument; communal property will not be looked after as carefully as private property.
Classical	• The last quarter of the eighteenth century and the beginning of the industrial revolution opened up vast possibilities of expansion to the recently established mode of industrial capitalism. • England was the most advanced capitalist country and its political structure had begun to change in accordance with the ideas of liberalism long before the French Revolution released its stimulus for the forces of liberalism everywhere.
Adam Smith (1723–1790)	• The champion of the *laissez-faire* approach. • He was under no illusion about the desire of individuals, including businessmen, to create a privileged position for themselves. But he nevertheless believed in the harmony of interest, because he thought that these privileged positions could only be maintained with state support without the intervention of government. • He believed in natural order; however, he did not doubt the compatibility of social harmony with the institution of private property. He also believed in the defence of the rich against the poor by the government and thought that property was the chief cause of authority and subordination.

	• Claimed that inherent natural order is superior to any order artificially created by mankind and the wise social organisation needs to act in harmony with the dictates of the natural order. • He stressed the point that the benefit of one should not conflict with the good of others. • He believed that labour is the main source of the wealth of the nation and this depends on the degree of productivity of labour and the amount of useful labour.
David Ricardo (1772–1823)	• *Theory of value and distribution*: the whole produce is divided among the three classes of the community, that the proportions of this division vary in different stages of society, that to determine the laws which regulate this distribution is the principal problem in political economy, and that hitherto there has been very little satisfactory information respecting the natural course of rent, profit and wages. • *Analysis of the causes of the value*: the discovery of the origin of the surplus product and consequent classification of different activities and classes of society and of various policies in relation to the production, accumulation and distribution of surplus.
Edmund Burke (1729–1797)	• He was in favour of non-interventionism, except for state finance. • He favoured a wealthy and financially independent Church. • He was also in favour of rights to property. • In his philosophy, working classes are capable of governing.
Friedrich von Gentz (1764–1832)	• He thought that self-interest was the main motive of human conduct for the individual and society. • Regarded the development of trade, industry and scientific agriculture as unnatural, though he could not deny their usefulness. • Tried to compromise by stressing the power of public finance in moulding the economic activities of the community as a whole. • In favour of indirect taxation as an instrument of state policy. • Excessive power assigned to the state is much in evidence in his theory of money.
Marxism	
Karl Marx (1818–1883)	• Man is a social producer of his means of livelihood. • Labour, viewed as a producer of use-value, is not the source of value; for this labour cannot be exercised without some natural material. • The form in which the proportional division of labour operates in a society where the inter-connection of social labour is manifested in the private exchange of the individual products of labour, is precisely the exchange value of these products. • Social production involves certain social relations, the quality of which depends upon the degree of development of the social productive powers.

	• In order to allow the expanding productive process, social change is essential. Therefore, political and legal institutions have to change and so will ideas. • Society can be classified on the basis of the factor of ownership of the material means of production. This division is the basis of the process of production and all other economic phenomena. Once economic relations have been established, the process of production itself makes them subject to change.
Modern Alfred Marshall (1842–1924)	• In Marshall's view, the forces behind supply and demand are: 1. Behind demand is marginal utility reflected in demand prices of buyers. 2. Behind supply is marginal effort and sacrifice reflected in the supply price. • The equilibrium between supply and demand can be considered as: 1. the market value equity of supply and demand, when the supply is assumed to be fixed. 2. the nominal values, which relate to short or long periods.
John Maynard Keynes (1883–1946)	• Keynes was concerned with aggregate income, consumption, savings and investment, rather than with the determination of individual prices which formed the core of the economic theory. • Regarding the supply and demand functions: 1. Effective demand is defined as the aggregate income which the entrepreneurs expect to receive, inclusive of the income which they will hand on to other factors of production, from the amount of current employment which they decide to give. 2. A supply function can be established, relating the aggregate supply price of the output obtained by employing a variable number of men with that number. • The level of income determines the volume of employment. Furthermore, it is determined jointly by the level of consumption and by the level of investment. • Functional relation of employment, consumption and investment gives a certain marginal propensity to consume. • The psychological attitude to liquidity and the psychological expectation of future yield from capital assets could differentiate the Keynes model in economic theory. • *Determinants of the volume of investment*: when someone invests they purchase the rights to the series of prospective net returns which they expect to obtain from selling, the output of the capital asset in which they have invested during the life of the asset (marginal efficiency of capital). • Emphasis of government intervention when aggregate demand is inadequate to create full employment. • The essentials of the capitalist system could be preserved without sacrificing full employment if government exercises proper control. • There are certain areas that should not be left to individual initiative.

	• State has to exercise a guiding influence on the propensity to consume partly through taxation schemes and partly by fixing the rate of interest. • He proposed a mixed economy in which investment is socialised but in which private self-interest will continue to function in all areas in which it is compatible with full employment. He regarded this as “the only practicable means of avoiding the destruction of existing economic forms in their entirety and as the condition of the successful functioning of individual initiative”.
Chicago School of Thought	
Frank Knight (1885–1946)	• Frank Knight is considered the leader of this school, who emphasised the theory of individual choice based on households as the production units that seek to maximise utility much as firms seek to maximise profit.
George Stigler (1910–1991)	• They were pioneers in promoting law and economics as the chief interdisciplinary fields of the social sciences by studying the effects of the assignment of property rights.
Milton Friedman (1912–2006)	• This school was interested in the analysis of growth in underdeveloped countries via market mechanisms. • They were pro-liberalisation and in favour of economic freedom as a means to achieve higher efficiency of the market.
James Buchanan (1918–)	• They emphasised the usefulness of sovereignty of consumers for maximising behaviour in their economic activity. • They stressed the point that in addition to the stock of physical capital and increasing the labour force, technical progress and human capital are also important. • They believed in the philosophy of profit maximisation as a response to the incentives given to them (labour force) in underdeveloped countries. • The market mechanism can stimulate efficiency and growth in an underdeveloped economy more effectively than the alternative policy of governmental planning as an instrument of economic development. • The necessity for encouraging the emergence of ‘entrepreneurial personalities’ in underdeveloped countries is a matter of special concern. • They urged government to facilitate private investment by supplying data not generally available to individuals. Education, free elections and nationwide media were regarded as being useful in opening up an otherwise closed society.

Source: Keynes (1936), Knight (1950), Polanyi *et al.* (1957), Walsh and Gram (1980), Lowry (1987), Betz (1988), Rima (1988; 1996).

Alternatives to privatisation

Privatisation is not the only way in which a state can divest itself of economic industries. Another method is corporatisation. This serves as an alternative to either full nationalisation or full privatisation, with features in common to both. It is in many ways the road that Iran has taken in the past, an issue which is discussed in later sections. However, it might be helpful to deal with the issue here for the sake of thoroughness in our analysis.

Corporatisation is, at its most basic level, the process of moving the control of industries from direct control by the government to control by government-owned corporations (personal communication).[1] Within this definition there is room for manoeuvre. The government's control may be by a specific bureaucracy with sovereign power or a bureaucracy controlled by other officials. The corporation may be manned by government officials or have private citizens in charge. The government may retain some control over the corporation or divest it all. In some cases, corporatisation may simply be another way for the government to maintain control. However, in general, corporatisation is usually a method of economic reform and it occurs either as an alternative to full privatisation, as a middle ground, or as a stepping stone to full privatisation.

The process of corporatisation involves converting the government bureaucratic departments which are in charge of the industry into public companies. Commercial boards of directors are then inserted in between the ministers of the government in charge of the industry and the managers of the corporation (personal communication).[2] In this way, there is a mixture of the government still having strategic control of the organisation, but the efficiencies of private enterprise can be taken advantage of and some of the inefficiencies of government oversight can be controlled. However, this process of corporatisation does not generally create the level of efficiency necessary to compete with truly private corporations without government subsidisation.

Privatisation in Practice

This section considers privatisation relative to the geographical area in which they occurred in the neighbouring countries of Pakistan, Azerbaijan, Kyrgyzstan and India, and then in the sectorally related countries of Egypt and China. India, Pakistan, Egypt and China, with large populations and very old public sector enterprises, have all followed import-substituting industrialisation, in which state enterprise has played a leading role, while Azerbaijan and Kyrgyzstan were liberated from a command economy after the break-up of the Soviet Union. In 1982, Kirkpatrick (1982) claimed that the difference between totalitarian and authoritarian states is precisely that authoritarian regimes may become more democratic and pluralistic because they leave civil society intact. In contrast, totalitarian regimes, such as those existing in Azerbaijan and Kyrgystan, cannot be expected to relax control and evolve into freer societies. Egypt is grouped in this study as one of the authoritarian regimes in the region.

Privatisation in the Russian experience, while appearing to be an obvious case for comparison because of its proximity, was left out of this study simply because of the nature of its attempts and methodology in "corporatisation". The Russian model adopted a national voucher system, carrying out the first phase of the programme with 750 bid reception centres in more than 85 regions in anticipation of giving 144 million Russians a chance to become, along with insiders, owners of enterprises (World Bank, 1995). Commencing in November 1991, by June 1994 some 12,000 to 14,000 medium-sized and large enterprises had been transferred to private ownership. The national voucher auction system was disbanded in 1994 due to the many objections leading to what was to become known as the "piratisation of Russia" (Goldman, 2003).

Geographically related cases

Iran's privatisation programme does not occur as an isolated case in the region. There are numerous examples among its neighbours (closer or more distant) in terms of policies, strategies, institutional settings, legal frameworks and implementation mechanisms. In order to contextualise, to a certain extent, the Iranian experience, we will now examine the cases of Pakistan, Azerbaijan, Kyrgyzstan and India. One has to acknowledge of course the limitations of the comparative approach and take account of the likely possibility that the privatisation policy of the Islamic Republic could be exclusively home-grown. The spiritual leader's pronouncement of 3 June 2006 on the general policies enshrined in Article 44 of the Constitutional Laws of June 2004 – pertaining to the specifics of divestitures of state-owned entities – illuminates the specificity of the Iranian privatisation policy. Currently, Pakistan and India are negotiating with Iran on the transfer of gas through a pipeline which has now been termed the 'peace pipeline', while Azerbaijan and Kyrgyzstan are interested in oil swap contracts through the Caspian Sea for delivery to the Persian Gulf. Hence the selection of these countries as case studies in the discussions below.

Pakistan

In Pakistan, the impetus for privatisation was derived largely from the negative experience of state-owned enterprises (SOEs) (Pakistani Privatisation Commission, 2004).[3] Since the early 1970s, as in most developing countries, Pakistan has relied on the public sector to operate virtually all infrastructure and financial services and many industrial units. The government had nationalised many businesses and had enlarged the areas where the private sector had been prevented from competing. In addition to a number of important industrial enterprises, the government owned and operated services in banking, energy, communications, infrastructure and transport. The following problems persisted, creating the incentive on the part of the government to privatise state-owned companies: a) mismanagement and overstaffing; b) inappropriate and costly investments; c) poor quality and coverage of services; d) substantial debt and fiscal losses; e) production and profits well below potential.

The actual privatisation policy of Pakistan commenced with the creation of the Privatisation Commission on 22 January 1991. Although its mandate was initially restricted to industrial transactions, by 1993 it had expanded to include power, oil and gas, transport (aviation, railways, ports and shipping), telecommunications, banking and insurance. Between January 1991 and June 2003, the Commission completed 132 transactions with a total value of Pakistani Rs101.027 billion (£940.455 billion at current exchange rates). Initially, the number of transactions was quite low (slightly more than ten transactions per year on average), while in the second phase the number soared (i.e. revenues of around £78.300 billion pounds per year). Obviously the companies sold were of high calibre and value. The distribution of sales among sectors was as follows: a) 43% in the banking and capital markets; b) 24% in the energy sector; c) 18% in the telecommunication industry and d) 15% in other sectors of the economy.

In terms of organisation, the Privatisation Commission has the right to initiate for the sale of any given enterprise. It approaches the owner (ministry) with the so-called summary of why divestiture is desirable and may also seek guidance on issues relating to such matters as pricing, restructuring, legal considerations and regulatory framework. Following this, the proposal is submitted to the cabinet or its subcommittee on privatisation for approval. If granted, the Privatisation Commission has the option of hiring external advisors to evaluate the value of the company and prepare for its restructuring and divestiture. During the whole process it remains in control and must approve each step. The privatisation method most often used is that of bidding, whereby the highest bidder gains the contract which is then submitted for approval by the Commission's board. In practice the privatisation process is a lengthy one, mainly to ensure transparency at each stage. On average, it takes about 18 months to close a major transaction, even when no major restructuring of the company is required.

Azerbaijan

Privatisation in Azerbaijan has taken a somewhat different path (IMF, 1995). The policy was initiated in the mid-1990s with the following objectives: a) the establishment of self-regulated economic entities based on private ownership and free competition; b) the restructuring of the national economy in accordance with market-economy requirements; c) the regeneration of all unused resources (unfinished and decommissioned industrial entities, real estate, the population's free resources, entrepreneurial potential, etc.); d) the mobilisation of investments, in particular foreign funds, for the national economy; and e) the improvement of social conditions and living standards. Right from the start, the Azeri policy-makers indicated their wish to open the economy to foreign capital, perhaps correctly assessing the inadequacy of domestic resources, and to allow property acquisition by overseas enterprises and individuals.

The central privatisation organ is the Azeri State Property Committee, which has no right of initiative for divestiture but is merely the selling agency, not the owner of state enterprises. Initiative belongs to the respective ministries, which are required to submit to the State Property Committee's recommendations relating to privatisation. Only then is the Committee enabled to work out the particular details in preparation for the deal giving due regard to the opinion of the State Committee of Antimonopoly Policy. From the date on which the privatisation deal is concluded, all powers delegated by the state to any ministry, body, association, local executive authority, etc. ceases to exist and the State Property Committee is charged with overseeing the implementation of all post-privatisation commitments by the new owner, if such supervision is specified in the contract.

Kyrgyzstan

A similar formula was adopted in Kyrgyzstan, which launched its privatisation programme in 1991 (World Bank, 1998). Since then there have been three distinct phases to the process. Phase I lasted from 1991 to 1993. It focused on the establishment of the legal framework and institutional infrastructure required to enable ownership of private property (other than land) and private enterprise. This period saw two principal approaches to the privatisation of companies: a) the privatisation of more than 3,400 small-scale enterprises in the retail trade, catering and services by means of cash auctions and b) a mass privatisation programme, whose purpose was to transfer a portion of state property to the citizens of Kyrgyzstan free of charge in compensation for their life-long contribution to the economy. Phase II lasted from 1993 to 1996 and began with the adoption of a new privatisation strategy which brought about one important change. This was the creation in May 1994 of the Enterprise Restructuring and Resolution Agency under the World Bank-supported Privatisation and Enterprise Sector Adjustment Credit. The agency started to deal with large state-owned loss-making enterprises that needed either radical restructuring to enable their sale, otherwise they would go into liquidation.

Phase III was inaugurated in February 1996 through a governmental decree. It allowed the usage of previous privatisation mechanisms, such as coupon auctions, cash auctions and competitive tendering for large investment projects. It is, however, distinguished from the previous periods by the nature of the companies involved – predominantly large-scale enterprises in mining, construction, transport and tourism, public utilities and infrastructure projects. During this phase a number of companies that had already been partially privatised, that is where the government continued to retain significant shareholdings, were assigned to full private ownership. Such companies were reviewed individually and, where possible, offered for sale by the State Property Fund. In Kyrgyzstan, the latter was not merely the seller of state-owned companies but also the owner following the transfer of shares to it by the respective ministries, a fact that presumably gave it more independence in its operation.

India

Privatisation in India was driven by the need to meet the increasing domestic demand in the economy and to mobilise additional capital for setting up the necessary infrastructure to support it (Mishra and Navin, 2002).[4] The privatisation environment was assessed as lacking in attributes that are fundamental to a market economy, such as open access, fair play and competition, as well as attributes that are essential for protecting the interests of the consumers. The attempts to create an environment suitable for private investment triggered a process of reform, restructuring and deregulation. In the wake of economic liberalisation in 1991, the initial response from the private sector was overwhelming. However, the government's policy towards the participation of non-state enterprise in infrastructure development was thought to be fundamentally flawed. For example, in the power sector the government opened up energy generation activity to private investment, while, it was argued, privatisation should have started with the energy distribution network.

Given these shortcomings, amendments to the privatisation policies became routine, a situation that created crippling uncertainty for potential buyers (Mishra and Navin, 2002).[5] No less discouraging was the lack of an effective institutional framework for privatisation to proceed smoothly. Rather than having a central authority, either independent or subordinate to the government, a number of agencies were created to deal with the different sectors of the economy. Over the past few years a fair number of agencies have emerged: the Central Electricity Regulatory Commission, the Telecom Regulatory Authority, the Insurance Regulatory Authority, the National Highway Authority, the Inland Waterways Authority, the Airport Authority and others. Beside their normal functions of advising respective ministries on policy-making, they have also been discharging many regulatory functions as well as those relating to the transferral of certain state economic responsibilities and assets to the private sector.

Sectorally related cases

Privatisation in the oil and gas industry is a topical issue, and not only for the countries where it represents a sizable chunk of the national economy. Where revenues from the exploitation of oil and gas reserves represent a significant contribution to the budget, the pressure on governments to modernise extraction facilities and streamline production inevitably grows. Moreover, where capacity is to keep up with technological developments around the world, the case for enlisting the participation and support of foreign investors becomes compelling (personal communication).[6] This has happened by and large under the mechanism of concessions or guaranteed exploitation periods/quotas. A number of relevant cases are considered below.

Egypt

Egypt's petroleum industry plays a key role in the economy. The country is an oil exporter and the industry is one of the main sources of foreign exchange. The upstream oil and gas industry generates 10% of GDP, with oil accounting for about 40% of total export revenues. Egypt's 1999 proven oil reserves estimates are given at about 3.5 billion barrels (Mbendy, 2005). The main areas of exploration for oil and gas are the Gulf of Suez, the Nile Delta, the Western Desert, the Sinai Peninsula and the Eastern Desert. The Gulf of Suez, despite decreasing output, remains the major source of oil. The decrease in production levels together with the increase in domestic oil consumption (associated with gradual industrial growth in Egypt) has engendered fears that by 2010 Egypt might become a net importer of oil. This fact has generated policies of attracting foreign investors with technological know-how to participate in the upgrading of the country's oil and gas industry. The government, through the state-controlled Egyptian General Petroleum Corporation (EGPC), has offered attractive exploration terms to overseas companies and this has helped to dramatically increase oil and gas exploration and discoveries in the past decade. For example, the production volumes of 1999 had doubled by 2003. Following the introduction of an IMF stabilisation programme in 1991, the Egyptian government began a reform of its fiscal policy and announced attractive incentives to foreign investment and exploration, and a commitment to a policy of privatisation. The improved incentives for exploration include the provision of larger exploration blocks, increased cost recovery allowances and extension of agreement periods (personal communication).[7]

There has been significant progress in privatisation, trade liberalisation and deregulation during the 1990s. This occurred mostly in the more pervasive state-owned sectors, such as telecommunications and utilities (e.g. the Egyptian Electricity Authority). The General Petroleum Corporation and Egypt Gas (the natural gas administration) was not considered for privatisation, apart from granting contracts for limited exploration, similar to the buy-back agreements employed in Iran. Traditionally in Egypt, this has taken the form of joint ventures with multinational companies, without a transfer of property rights. After 1973, the joint venture scheme was replaced by production-sharing agreements, again with no transfer of property rights from the state to private companies. While privatisation in the other sectors of the economy has slowed down due to the large burden of debts of state-owned enterprises, privatisation of the oil industry is not on the table because of its strategic importance (World Bank, 1997; Ibrahim and Ibrahim, 2003).

China

Unlike Egypt, the importance of the oil sector in China is not due to its export capability, but to domestic demand. In late 2003, China surpassed Japan in becoming the world's second largest petroleum consumer after the United States. Associated with rapid economic growth, as argued by Logan (2005), internal Chinese demand grew by 15% to 6.37 million barrels per day during 2004, about one-third the level in the US. In comparison, domestic oil output in China has grown only very slowly over the past five years. As a result, imports of crude oil grew rapidly in 2003 and 2004, increasing by nearly 75% over the course of the two years. This has generated a need for the modernisation of the oil and gas sectors and was met by governmental investment as well as through the involvement of overseas companies.

With China's accession to the World Trade Organisation, the government made several important commitments regarding the energy and oil industry. It agreed to gradually open the crude and refined oil sectors to private traders and to cut back its state monopoly over oil trading by handing over 4 million tons of oil products and 10% of crude imports to the private sector. After three years of its accession to the WTO, China was to deregulate retail oil distribution by enabling foreign investors to own at least 30 petrol stations each. It pledged to open its wholesale market five years after accession. By December 2004, China had removed import quota management from the three state-owned oil companies: China National Petroleum (CNPC), China Petroleum Company (Sinopec) and China National Offshore Oil (Anon, 2004a). In October, the French oil giant Total was allowed to enter into a joint venture agreement with Sinopec to develop petrol stations in the country, and in November, British Petroleum signed with Sinopec and CNPC to build petrol stations.

In terms of privatisation (Majumder, 2003), a broad sectoral policy was announced at the 15th Central Committee meeting of the Chinese Communist Party in 1999. Its leadership reaffirmed the earlier stance of Jiang Zemin and approved the programme, thus eliminating the political constraints over the issue. The concrete policy on privatisation was announced in the third Five-Year Plan in 2001. According to this, the economy was divided into three areas: a) areas of highest national priority which could not be relinquished from full governmental control; b) areas which need joint participation and would remain partially controlled; c) areas in which complete privatisation would be allowed. Defence, for example, fell in the first group, while public services and national priority sectors, such as power, waterworks, gas, onshore oil and transport fell in the second group of state-controlled development with private participation.

At the initial stage, China faced internal squabbling over privatisation. After the congress meeting in 1999, disputes between the reformists and conservatives arose. With the former winning the upper hand, however, local governments started to sell small- and medium-sized state-owned enterprises. The privatisation strategy which was adopted for large companies was two-pronged. In the first phase, they would be converted into joint stock companies/cooperatives and, in the second, the government-held shares would be partially divested to private enterprises, including managers and employees (personal communication).[8] Most probably as a result of its privatisation programme, China, in 2002, was the world's biggest beneficiary of foreign direct investment. In the oil sector, however, governmental control has been retained owing to its overwhelming importance to the future economic growth of the country.

Rather than looking to sell assets of the domestic oil industry, the Chinese government has instead sought to diversify its supply portfolio by acquiring them overseas. Over the past decade, Chinese oil imports have come from an increasingly wider network of suppliers. In 1993, for example, almost all of them came from Indonesia, Oman and Yemen. By 2004, Saudi Arabia was China's largest supplier accounting for 14% of imports, with Oman, Angola, Iran, Russia, Vietnam and Yemen together supplying another 60% (personal communication).[9] The third Five-Year Plan (2001–05) called for the construction and use of strategic petroleum reserves by the year 2005. Construction has begun at one of four sites earmarked for the storage of government-owned supplies; Chinese officials plan to gradually fill the depots with up to

100 million barrels of storage by 2008 (roughly equivalent to 35 days of imports). Within this strategy, Chinese state-owned oil companies have accelerated their hunt for overseas oil assets. Until recently, they had been most comfortable operating in locations not dominated by the world major oil giants. Sudan, Angola and Iran are pertinent examples. Currently, over half of the Chinese overseas oil production comes from Sudan. Most recently, outward activity has also gathered speed in other areas, including Russia, Kazakhstan, Ecuador, Australia, Indonesia and Saudi Arabia. The impressions shared by Logan (2005) is that Chinese companies seem to be improving their ability to purchase assets without overpaying, as earlier reports had suggested, and to be developing more effective business skills.

Leading the drive among Chinese state-owned companies, the CNPC claims to have petroleum assets in 30 countries. It plans to spend $18 billion in overseas oil and gas development between now and 2020 (Anon, 2005a). Most of CNPC's overseas production currently comes from Sudan, Kazakhstan and Indonesia. Many speculated that CNPC would take a share in the restructured assets of Yukos, while rumours in late January 2005 foresaw a $6 billion 'loan' by CNPC to Rosneft for long-term oil purchases, but no equity investment. By comparison to CNPC, Sinopec is new to international investment in the oil sector. It has recently started pumping smaller quantities of equity oil from operations in Yemen and Azerbaijan. The largest of Sinopec's overseas contracts was signed with the Iranian government in 2004 and envisaged an investment of $70 billion over 25 years to purchase cargoes and to participate in upstream oil activities.

Relevance to Iran

The two sets of precedents of privatisation are relevant to Iran, one comprising countries geographically close to the country, the other comprising countries that share sectors for potential privatisation. Although the first category, namely Pakistan, Azerbaijan, Kyrgystan and India, are not necessarily similar to Iran in political, economic and, specifically, privatisation terms they are nonetheless pertinent and need to be considered. While acknowledging their differences, there is also good reason to believe that the common historical and geographical features of the region have produced certain common denominators in this group of countries. The second set of suitable cases, Egypt and China, have oil and gas, like Iran. These are countries where the revenues from the exploitation of these resources represent significant contributions to their budgets, a fact which explains the increasing pressure on their governments to modernise extraction facilities and streamline production. However, if the native capacity to keep up with technological developments around the world is lacking, there is a strong case for enlisting the participation and support of foreign investors. This has often led to the use of the mechanism of concessions or guaranteed exploitation periods/quotas.

Within both of these sets of case studies, it is possible to point to three important considerations for privatisation in Iran. The first is the need for an impetus to privatise, the second is that there are a number of different models of privatisation that recur across different countries, and the third is that certain lessons can be learned from the experiences of these precedents.

Impetus to privatise

One of the relevant findings, noted in the case of Pakistan, is that in some so-called developing countries, one of the primary factors that led to privatisation was the disappointing experiences of state-owned enterprises. The reasons behind their poor performance include the following factors:

- Mismanagement and overstaffing;
- inappropriate and costly investments;
- poor quality and coverage of services;
- high debt and fiscal losses;
- production and profits well below potential.

These shortcomings have bedevilled Iran and are bound to give impetus to further and faster privatisation.

Another similar set of factors relate to inadequate market conditions. These include:

- Lack of open access;
- limitations on "fair play" and competition;
- lack of measures to protect the interests of the consumers.

In India, as in other developing economies, attempts were made to correct these failings in order to make Indian goods and services competitive in world markets. One of the steps taken was to privatise a number of industries.

Models for Iran

The number of different models of privatisation set out above may possibly provide the groundwork for a new model for Iran and offer lessons in what mistakes to avoid in its privatisation decisions.

Lessons for Iran

One notable lesson learned relates to the duration of the privatisation process. For instance, it was noted that in the Pakistani model the privatisation process is lengthy, mainly to meet the requirement of transparency. Eighteen months is too long a period considering that it applies even when no major restructuring of the company is required. The lesson for Iran is to expedite implementation, otherwise there would be infavourable repercussions on private investment.

Another lesson learned is that governments may simply be incorrect in laying down the sequence for privatisation. For instance, in the Indian case study, it is argued that the government's policy towards private participation in its infrastructure development was fundamentally flawed in that it opened up energy production activity for private investment rather than the energy distribution network, which should have taken precedence. As a result of this and similar misjudgements, policies were constantly amended, creating a great deal of uncertainty for potential buyers. This is relevant to Iran because it is often assumed that the government carries out fairly correct analyses of the situation and that its policies would reflect them. This is not always true, however, and the government would be well advised to heed the experience of India.

A further lesson to remember is that not all industries are suitable for privatisation. In general, industries which are considered to be strategic are generally less open to privatisation. In the case of China, it was shown that this consideration had led to a three-level set of privatisation formulas. In the Egyptian case the General Petroleum Corporation and Egypt Gas were not considered for privatisation; only contracts for exploration of limited duration were granted. The government judged the oil industry to be of strategic importance which other industries, such as the telecommunications industry, lacked. Similarly, in the oil sector in China, governmental control remained unaltered due to its overwhelming importance for the economic growth of the country. In this respect, Iran is sure to take this view into account, especially as it has always regarded its oil and gas industry as eminently strategic.

Thus, the nature of the industries must be considered in any privatisation decision. The experiences of other states provide a precedent for the oil industry which cannot be treated as other less strategic sectors. However, one should not take this to mean that, because an industry has a strategic character, there is no justification for its privatisation. Such a rigid attitude would never make for change. The Chinese case is a good example of a compromise solution. Despite the fact that the government has retained more control over the oil industry than over other industries, transformations have occurred. While, as noted earlier, the example of China is in some ways very different from the Iranian case, it shows that there is always the possibility for other models to emerge which bring about economic change and make industries more competitive. Another lesson learned is that political resistance to privatisation is the norm in some countries, not the exception to the rule as in others. Even in China, which has a centralised government structure and an extremely centralised economic system (although it is on the way to becoming less so), there were disputes, after the congress meeting in 1999, between reformists and conservatives. With the former gaining the upper hand, local governments were able to sell small- and medium-sized state-owned enterprises.

Controversy has dogged Iranian attempts at privatisation, and the outcome of the 2005 presidential election in which Ahmadinejad came to power has deepened it further. Yet, there is light at the end of the tunnel as controversy and debate also mean that the issue is discussed and evaluated. So, while the core of the Iranian government opposes the privatisation of the oil and gas industry, the debate goes on and, out of this debate, a compromise might emerge in which parts of the industry might be considered for privatisation.

NOTES

1 Various discussions with Abdullah P. Hosseini, Head of Iran's Privatisation Organisation (IPO), during August 2004.
2 Discussions with Seyed Ahmad Mir Motahari, Head of IPO, December 2003.
3 This section is based on official information published by the Pakistani Privatisation Commission (2004). Information on privatisation available on-line at www.privatisation.gov.pk
4 See also www.indiacore.com.
5 See also www.indiacore.com.
6 Personal communication. Discussion with Mohammad Alipour-Jeddi, Head of Petroleum Market Analysis Department, OPEC, NIOC Headquarters, Tehran, December 2005.
7 Personal communication. Discussions with Seyed Ahmad Mir Motahari, former Head of Iran's Privatisation Organisation (IPO), February 2004.
8 Personal communication. Discussions with Seyed Ahmad Mir Motahari, former Head of Iran's Privatisation Organisation (IPO), February 2004.
9 Ibid.

2

POLITICS AND ECONOMICS DURING THE PAHLAVI REGIME

Introduction

This chapter begins with a summary of Iran's modern history to lay the groundwork for an account of the most important trends and features of Iranian political and social life, and relates them to the specific oil policies of Iranian governments on privatisation and nationalisation. The chapter also analyses the roles played in the past by internal and external events and, not least, by key figures inside and outside Iran, all of whom brought about significant changes in the country's political and economic life that led to the making of today's government.

The chapter outlines the reasons behind the Iranian government's decision to nationalise the oil industry and then partially privatise it. The issue is dealt with within a broader discussion of the role of internal and external factors (economic, political, social and cultural) that determined the evolution of the Iranian state. The narrative serves to contextualise the analysis of post-revolution Iranian politics and economics and runs from World War II – regarded as the beginning of modern Iranian history – through to the Iranian revolution in 1979, which may be seen as having shaped the features of the Iranian government up to the present. It argues that the pre-revolution period was characterised by centralisation of power, nationalistic discourse reflected in the policies relating to the oil industry and clientelistic relationships between the authorities and the business community. A failed constitutional reform, due to both internal political factionalism and the centralisation of power in the hands of the Shah, hindered political and socio-economic reform in Iran. A contradictory relationship with Western countries (dependence and independence) played an important role in the Iranian political and decisionmaking process. The relationship was marred by distorted perceptions (Jervis, 1976), but also nurtured by the inadequate fit between the economic and political interests of the West (specifically of the United States) and of Iranian society and government.

Politics during the Pahlavi Regime

The aspirations of the founder of the Pahlavi dynasty, Reza Shah, were focused on the development of the infrastructure of the country. This involved the creation of a viable railway system connecting the north to the southern ports, as well as a network of roads across the Iranian plateau in anticipation of making the country independent of foreign power politics. Along with this endeavour, the politics of the country leading up to the time of his son, Mohammad Reza Shah, consisted of modernising the military and financial institutions, and the creation of a functioning civil service to cope with the anticipated income from the export of oil. This section examines the role of Mohammad Reza Shah and his plans, which aimed, firstly, to achieve rapid growth and high employment, and secondly, to achieve a degree of economic diversification which would result in the more equal distribution of wealth and the creation of a middle class.

Political and economic modernisation, 1906–53

From the onset of its modern history, Iran wanted to modernise its political and economic systems. These processes were accompanied by an important nationalist discourse, which fitted well with the European discourse on the rights of self-determination (Katouzian, 1981:130–2). The first stage of the modernisation process can be traced back to the Constitutional Revolution of 1905–6, which brought about an elected assembly. But it was during the second stage – initiated in 1925 by Reza Shah Pahlavi, a "ruthless nationalist" who had seized the throne (Milani, 1997:80) – that the foundations of the Iranian oil industry were laid. It was during his rule that Iran became a key oil exporter.[1]

Following Reza Shah's enforced abdication in 1941 in favour of his son, Mohammad Reza Shah, Iran experienced decentralised political power that was split between a multitude of political actors between 1941 and 1953. Mohammad Reza Shah's court was only one centre of power amongst many, with the Shah himself having little power. This was one of the few periods in Iranian history when there were many political actors, ranging from political parties to student associations, labour unions and activists promising political freedom. Milani describes the situation as follows:

> The level of political participation reached a new zenith. New ideological parties were formed, recruiting people from all social

> classes. Activists published new newspapers, professionals and students formed associations, labour organisers revitalised their unions, and political debate became fashionable and free (Milani, 1997:80).

Besides such plurality of politically active players, factionalism was also a feature of the Iranian political scene. Political factionalism is characterised as the inability of political players to form coherent and long-lasting coalitions in order to implement policy platforms. It is argued that when political parties offering properly articulated positions are suppressed so that they are unable to form a broad coalition to attract mass support, a situation arises where competing economic and political interests pursue power in a clandestine fashion, often by forging unstable alliances between themselves or by attempting to gain influence or favour with the executive power. This kind of political suppression characterised the period leading up to the Islamic Revolution.

Factionalism was reflected in hundreds of fragmented and fragmenting parties (Shafaq, cited in Abrahamian, 1982).[2] Individualism seemed to guide the behaviour of political players in a way that made Shafaq notice that deputies behaved as if they were the sworn enemies of the ministers. A major consequence was an unstable political environment.[3]

Abrahamian points out, however, that despite the variety of actors, one can safely argue that, overall, a relatively small elite still retained effective control over what political measures to adopt, and this was due to the lack of an educated middle class politically independent of the palace. He writes: "Of the 148 ministers, only fifteen were salaried personnel and modern, educated professionals with roots in the middle classes and without links to the palace" (Abrahamian, 1982:170).

Nationalisation of the oil industry

When a political actor could rally the support of the middle class and the support of the ulamas,[4] power could be gained and exercised. A case in point was the election of Mohammad Mossadeq as Prime Minister in May 1951. Commenting on his success, Milani writes:

> Although he and his cohorts constituted a very small faction in the Majles, they relied on public pressure, mass demonstrations by their supporters, and an alliance with several leading ulama to pressure the landlord-dominated Majles to approve their innovative nationalisation plan (Milani, 1997:80).

In spite of his own personal background, which placed him amongst the landowning conservative elite, Mossadeq and his National Front Party put forward a progressive programme which included the reform of the electoral system to the 1943 *Majlis*. It was aimed at further democratising the country and removing the Shah's control of the military to prevent a rerun of Reza Shah's dictatorship. It supported electoral reform to increase the voice of the professional middle class at the expense of the landed families. In terms of foreign policy, he believed that Iran should have avoided granting exclusive economic concessions to individual foreign powers (designed to buy favours) and that Iran should pursue a non-aligned course. One measure that directly concerns this study refers to the nationalisation of the main economic sectors.

By 1950, the pre-eminence of the oil industry within the Iranian economy meant that oil nationalisation was a major target. In the same year of his election, Mossadeq nationalised Iranian oil – at the expense of the British oil giant the Anglo-Iranian Oil Company (AIOC) (personal communication).[5] The fact that Mossadeq's reforms led to his downfall speaks volumes about the political Iranian environment, as well as the role of international players in Iranian politics at that time. Conservative groups, such as the landowners and army officers, were still influential, while the Shah remained reluctant to embrace institutional reform that could result in the loss of his control. The Shah's distaste for Mossadeq's reforms coincided with that of major international actors, such as the US and the UK (personal communication).[6] The latter responded to the nationalisation by imposing an oil embargo on Iranian oil exports. Then, with assistance from the US and British secret services, the Iranian military organised a successful coup against the Prime Minister in August 1953. Mossadeq and other National Front leaders were arrested and imprisoned. Mohammad Reza regained full control over the government.

One can conclude that in the first stages of modernisation, Iran made salutary attempts to democratise the country by creating a favourable environment for the political expression of various groups. These measures were, however, countered by the existence of a powerful conservative elite (landowners, military) loyal to the monarch, Mohammad Reza. This translated into resistance to institutional change and political reform. In other words, pluralism was allowed to exist so long as it did not radically change the socio-economic and political status quo. Factionalism, as a feature of Iranian political life, also contributed to the failure of the progressive programmes put forward by Mossadeq. Just as important, international actors had effective constraining mechanisms to influence Iranian domestic political life. The narrative so far has pointed to some important factors that influenced Iranian decisionmaking and the shape of economic policies with respect to the status of the oil and gas industry in particular. An analysis of the Shah's absolutist rule would suggest that some of the already existing tendencies or features would be consolidated.

Iranian perception of the West

When discussing the determinants of Iran's economic policies regarding privatisation, one needs to understand the perceptions of Iranian politicians and the wider public vis-à-vis foreign players, since privatisation and the opening up of the Iranian economy creates opportunities for the involvement of the latter. Thus, a discourse about Iran's external relations is necessary for the purpose of this study.

Iran's relations with its neighbours and powerful Western nations have always been difficult because of the country's strategic geographical location, its significant oil wealth and its weakness in comparison to the major Western powers. Although Iran was never directly colonised in the modern era, the extent of Western interference has been significant. Moreover, as Halliday argues, this atypical "semi-colonisation" of Iran meant that it never experienced a clear break separating a dependent past from an independent present and future.

> The patterns of semi-colonial control used by Britain, Tsarist Russia and later the US – influencing ministers, fostering dissension in the provinces and suborning the military – were precisely those most likely to engender a conspiracy mentality among Iranians (Halliday, 2003:65).

The Iranian people have good reason to view foreign powers with suspicion, given the extent of their interference in the country's internal affairs over many decades, specifically in regard to Iran's political system and policies concerning the oil industry. Instances of British, Russian/Soviet and American interference include: the secret Anglo-Russian Agreement (a plan to partition the country into zones of influence) exposed in 1907; the 1915 agreement between the same two countries (known as the Constantinople Protocol) that aimed to dismember Iran; the 1919 Iran–British Agreement that forced Iran to employ British advisors at Iran's expense; the involvement of British military officers in the 1921 coup that installed the Shah's father as ruler; the 1933 Oil Agreement which Iran signed under British pressure; the 1941 Anglo-Soviet invasion and, of course, the British/US-sponsored coup against Mossadeq in 1953. Unsurprisingly,

> . . . there has seldom been any significant (major as well as minor) political event, which . . . a large section of the Iranian public has not believed to have been planned and executed by foreign powers through their Iranian agents (Katouzian, 2003:118).[7]

It is interesting to note that the Iranians' view of outsiders as conspirators and schemers only emerged during the modernisation stages. As Katouzian points out:

> Up to a century and a half ago hardly anyone attributed perceived injustice to the machinations of western powers or their paid Iranian agents. But since then, because of the growing weakness of Iran vis-à-vis imperial powers, almost all injustice emanating from arbitrary states was attributed to imperialism (Katouzian, 2003:29).

One may add, however, that in addition to the events justifying Iran's mistrust of foreign powers, a blurred image of them in Iranian eyes also played an important part. That distorted view has doubtless contributed to Iran's prejudice against the West, one that is shared by the public and policymakers alike.

The Reign of Mohammad Reza Shah

On 25 August 1941, the 21-year-old son of Reza Shah, Mohammad Reza Shah Pahlavi, was installed on the Peacock Throne by the Allies. He inherited a country which was occupied by British and American forces in the south and Russian forces in the north. The German invasion of Russia, in June 1941, brought Stalin closer to Churchill. By this time, Reza Shah's regime had seen the completion of the Trans-Iranian Railway linking the ports of the Persian Gulf to the Caspian Sea. Consequently, it was not surprising that the Allied forces occupying Iran would want to use the networks to supply the Russian troops in their fight against the German army in Russia. At that time the British navy had started to convert from coal-burning engines to fuel-driven propulsion.

The new Shah remained mostly in the background at the beginning of his reign. His ambition was to consolidate his internal base of support, get closer to the American administration, and to start to behave as more than just a figurehead. Perhaps his biggest ambition was to build a modern Iran strong enough to compete with and stand up to the authoritarianism of European powers which wanted to dictate how he should run the country. It is perhaps ironic to see that, at the end of his rule in 1978, the revolutionary crowds were chanting slogans against him for having sold Iran out to foreign powers.

Central power and arbitrary rule

The return of the Shah from abroad after the fall of Mossadeq represented the beginning of a strongly centralised military dictatorship. In 1953 any opposition foci were eliminated, including the leading leftist Tudeh Party – the military executed 40 of its officials over the next four years (Abrahamian, 1982:280). In 1964, Ayatollah Ruhollah Khomeini was expelled to Turkey by the government for his fierce opposition to the regime and his personal attack on the Shah.

Sensitive to external forces, such as the United States which pressed for liberal reforms, the Shah attempted to reform the political system in 1958. The two-party system was abandoned in 1964 when the parties were dissolved and the Iran Noveen party was formed in their place (Milani, 1997:80). In 1975, he created the Rastakhiz (Resurgence) Party, "a mass organisation to which all of Iran's bureaucrats and persons of influence and importance were first encouraged, and then forced, to belong" (ibid.). However, all parties were "totally subservient to the monarch, and lacking any integrity" and, as a result, "failed to attract any significant portion of the population" (ibid.). The creation of government-sponsored trade unions by the secret service and the integrated traditional systems of local governance – such as village headmen – were also part of the same attempt to superficially diversify the political scene (Owen, 2000:93). However, these cosmetic changes failed to achieve his goal because of his overbearing desire to exercise full control by means of undemocratic measures.

Clientelism became a feature of the political system alongside the centralisation of power through repression. This system of bestowing privileges on the conservative property-owning classes had existed even before the Shah's authoritarian rule had taken root. Katouzian explains:

"The state has not been representative of the higher social classes; on the contrary, they have been its clients by virtues of the privileges it has bestowed upon them" (Katouzian, 2003:29).

To this, one must also add a long-standing disjunction between the wider society and the state. The relationship between the state and society was considerably different from that established in Western states. The idea of a social contract between the state and society was never the driving force in the organisation of Iranian society. Thus, the Iranian state lacked legitimacy and its decisions were not always based on law. Inevitably, such arbitary rule led to political instability. Katouzian argues that the consistent motive behind Iranian revolutions and revolts is the overthrow of an unjust ruler and the establishment of representative government. Because the ruler did not represent any particular social group or groups, when revolts occured, society coalesced rather than fragmented.

> The constitutional revolution of 1905–9 and the revolution of February 1979 . . . shared the basic features of traditional Iranian revolts inasmuch as they were aimed at the overthrow of the state – indeed the person of the ruler – at all costs, they had the support of the entire political society, and they were not resisted by any social class (Katouzian, 2003:31).

As mentioned earlier in this chapter, factionalism was specific to Iranian political life, and made a great impact on post-revolutionary Iran. Yet, oddly enough, this feature emerged to the surface even during the authoritarian regime of the Shah. The Rastakhiz party developed into an embryonic pluralist system but it soon spawned two "wings", which put slightly different slants on the official position. Zonis describes the situation as follows:

> It began to offer the possibility for communication, not merely from the top down, but also from the less to the more powerful. 'Wings' of the party were created . . . offering slightly different and legitimate perspectives. Increasing numbers of the politically literate began to believe that the formation of these wings offered, finally, hope for political democratization (Zonis, 1983:598).

This trend created some hope for democratisation but the structural authoritarian regime dashed it by banning competition between the two wings of the party in the Majlis, shutting down the party in 1977. This was a serious mistake:

> By the end of 1977 . . . just when revolutionary fervour elsewhere in the society was being translated into revolutionary action – the only national organisation capable of providing political support for the regime was undone . . . The economic and political processes which resulted in feelings of betrayal and frustration were significant sources of the revolutionary energy that culminated in the overthrow of the Shah (Zonis, 1983:599).

The repression of all independent sources of power or ideas was the guiding principle of the regime.

External Relations

Relations with external powers continued to follow the same trajectory during the Shah's time as it had done before. Actual interference by Western governments, on the one hand, and the need to find a scapegoat for the structural problems of Iranian economy and political life, on the other, aggravated the public's anti-Western stance (personal communication).[8]

At crucial moments, foreign intervention interacted with domestic factors to determine the scope and direction of Iran's political development. British and US involvement in the 1953 coup against the nationalist politician Mohammad Mossadeq (for his nationalisation of British oil interests in the country) put an end to an emerging pluralist political culture (Zonis, 1983:603; Katouzian, 2003:119). Later, pressure from the United States under the Kennedy administration to undertake liberalising reforms led the Shah to implement his "White Revolution", a general programme of policy reforms focusing in particular on land redistribution. The policy backfired because it alienated the Shah's supporters amongst the large landowners and paradoxically upset even those who approved of the policy because of their aversion to foreign interference. Some commentators also argue that the Carter administration's calls for liberalisation undermined the Shah's ability to combat the growing insurrection from 1977 onwards (Owen, 2000:92).[9]

The Shah's complex relationship with the West in general, and the United States in particular, is widely perceived as being of central importance in explaining both the evolution of his rule and his eventual downfall. It has been argued that "during the 1950s and 1960s, the United States – either through direct suggestion or judicious innuendo – could influence the Shah's behaviour and make him carry out American-inspired policies in Iran" (Zonis, 1983:604). Although the Shah's government also enjoyed close relations with non-Western countries,[10] his dependency on the United States coupled with the failure to reform his repressive system, exacerbated the anti-Western discourse within the opposition and society. Thus, nationalism, and anti-US rhetoric proved to be critical ingredients in the appeal of the anti-Shah forces in the late 1970s. The revolution "was felt to be a nationalist movement against the political, economic and cultural influence of the West, and of the United States in particular" (Halliday, 2003:64).

There were certainly imperial ambitions on the part of Western powers concerning Iran and these were played out in the internal politics of the country. Nonetheless, the Shah's arbitrary rule could not be blamed completely on the imperialists, but the public liked to think so. In Katouzian's words: "Imperialism was real enough but it did not create the arbitrary system of government in Iran, although it clearly did what in the circumstances it thought would serve its interest best" (Katouzian, 2003:29).

All the same, whether or not imperialism was to blame, what was more important was the deep-rooted anti-Western perception that influenced Iranian politics and policymaking. This factor comes up repeatedly in the analysis of subsequent eras of Iranian politics and is counted as one of the major determinants of Iranian policymaking and, as is discussed in more detail later, as one of the major determinants of Iranian privatisation policy. Interestingly, this perception does not apply to all the other Western powers. For example, despite Britain's involvement in Iranian politics in its recent history, the sentiment towards the West relates specifically to Iran's interaction with the United States. Consequently, the policies of the Iranian government were targeted in relation to the political and economic effects they had on the United States (personal communication).[11] The history of external interference helped to foster the "conspiracy mentality" which Khomeini played on (Halliday, 2003:63).[12] It has influenced not only the overthrow of the Shah but also the subsequent policymaking (i.e. the possible course of economic and political policies) and the attitude towards foreign investors and privatisation.

Economics under the Shah

Shah Mohammad Reza Pahlavi acceded to the throne in 1941 amidst foreign occupation, economic disintegration and multidirectional attacks on his father's 20-year rule (Zonis, 1971). His main focus was to consolidate popular support and admiration by stimulating rapid growth, high employment and providing comprehensive public welfare services as a means of achieving equitable distribution of wealth (Amuzegar, 1999). The reform and economic programmes throughout his rule were aimed at winning the backing of the masses and, to this end, developed the non-oil-based industries which would also make Iran not too dependant on the revenue from the sale of oil and petrochemical products.

His economic diversification started with the introduction of land reform during 1961–1962. He called the programme the "White Revolution" (Enqelabe-e Sefid) designed to gain the support of rural people and turning farmers into stakeholders. The programme attracted the support of the American administration and his American-backed Prime Minister, Ali Amini (Saikal, 1980). The plan, which initially looked noble in its intention, namely to close the gap between the big landowners, the political elite, the small number of industrialists, on the one hand, and the mainly agrarian population, on the other, turned out to be ill-conceived and short-lived. Despite extensive American aid and increasing oil revenue, by the mid-1960s the Iranian economy had not improved. In fact, the inequality in incomes was widening, and the urban population's standard of living, not to mention that of the rural population, did not improve as expected. By mid-1970, corruption had become widespread throughout the administration and Iran's economic growth continued to remain one of the slowest of the developing countries, with illiteracy running at the astounding rate of 85%.

The distribution of shares in assembly plants, which were touted to be major production units, attracted farmers from the rural communities into the urban population centres of the country. Inflation, due to shortages, rose to all-time highs during the years preceding the Islamic Revolution. The lack of research and development in the production units and industries erected by those close to the Shah and his immediate family (ones that were held up by the Shah and his cronies as symbols of progress), meant that they were ill-equipped to compete with the vast volume of imports.

In the following sections, this study examines other aspects of the policies that the Shah adopted to develop the foundations for a sound economy. However, as we shall see, they caused yet further alienation from the very people to whom he intended to bring prosperity and living standards to equal those in Europe.

Denationalisation of the oil industry

Oil has been the cornerstone of the Iranian economy and consequently the most important economic decisions concerned the regulation of the industry. However, due to the importance of the resource to the industrialised world in general, and the specific links with British and American companies in particular, these decisions were never purely economic. Politics was an important part of the decisionmaking game (personal communication).[13]

In the history of the oil industry, the Shah's decision to reverse the nationalisation of this sector is an important one. His signing of the Consortium Agreement – based on a 50-50 profit-sharing principle – which Mossadeq had rejected, signified that these decisions were not impossible to make, but were largely dependent on the broader political environment. One must remember that the Shah's regime was heavily supported by the United States because of Iran being a stable source of cheap oil for the West. However, the dependence of the Iranian economy on oil exports, especially in the 1970s, led the country to become a dependent capitalist state (Ehteshami, 1995).

Oil allowed for significant development within Iran. Increasing global demand for oil was the key motivating force behind Iran's economic development, providing both the funding and the rationale for much of the country's rapid industrialisation under the Shah. Total oil revenues between 1964 and 1974 stood at US$13 billion. Revenues between 1975 and 1977 alone were almost triple this figure (Abrahamian, 1982:427). This growth in revenue was fuelled by the sharp increase in oil prices in 1973-4.

Relations between the state and the capitalist class

It was obvious to the Shah that control of the oil industry was key to controlling the country's most important source of revenue. Thus, despite the agreements with the Western companies, the Shah tried to use his position of control to further consolidate his regime at home (personal communication).[14] Clientelism remained a key feature as previously, not only in the political sphere but also the economic one. The state's control of the oil resources fostered an unhealthy relationship between capitalists and the state with the former dependent on the latter (Nazmi, 1989:41–2). The Shah instituted an intensive capitalist development in Iran which involved an expanded role of the state in the economy by being both the initiator and the interventionist. Private capital accumulation was possible with the help of the state, and there were entrepreneurs keen to engage in a few highly profitable industries, such as construction. However, they were dependent on state "hand-outs", since the state which had a monopoly control over oil was the "fountainhead of capital accumulation" (Ehteshami, 1995:78). This monopolistic control further contributed to the autonomy and supremacy of the state in relation to society.[15]

Control over oil revenue allowed the Shah to further distance himself from society. As the oil revenue grew from the mid-1970s, the Shah was able to increase the country's military budget and reduce its dependence on foreign assistance. In turn, this permitted further centralisation of power and, as mentioned previously, ultimately distanced his rule from society. He began to focus almost exclusively on foreign, economic and military issues, to the detriment of internal political issues (Nasri, 1983:623).[16] In particular, much of the state's money was spent on military purchases. Under the Shah, Iran's armed forces grew exponentially (Nazmi, 1989:52). It has been noted that for every $1 earned in oil exports to the United States, Iran imported $2 worth of military equipment during the 1970s.[17] Moreover, genuine private sector development was undermined because the members of the capitalist class, created and nurtured by the Shah's development policies, were able to compete unfairly against other, less favoured, entrepreneurs. In Katouzian's words: "[These] were people who, often with relatively little capital of their own, acted as little more than agents of the state empowered to use public funds, and other privileges, for supplying assembled [products]" (Katouzian, 1981:278).

This practice brought about "modernisation in form", but it actually "spread . . . underdevelopment to the very pillars of the economy" (Nazmi, 1989:42). It proved that economic measures (despite relaxed relationships with foreign companies) cannot determine genuine development because of the continuous political interference by the state. Clientelism and authoritarian rule undermined healthy economic expansion and eventually led to the state becoming the centre of the Iranian economy. By the 1970s, the state was providing 70% of the total investment funds in the economy. The private sector was largely dominated by only 45 families who controlled 85% of the larger privately held firms (Ehteshami, 1995:81). One could say that the positive aspect of a capitalist layer in society was overshadowed by the economic distortions engendered by clientelism and arbitrary rule. In addition, the focus on military expenditure drained resources that could have been allocated to other sectors of the economy.

The state of the economy

Income distribution aside, the Iranian economy was in good shape. In fact, judging from the macroeconomic aggregates, economic policy seemed successful under the Shah. Real output grew by around 7.5% per year on average during the 1960s. Between 1965 and 1973, average real output growth stood at more than 10%. Indicators such as infant mortality and life expectancy also improved (Nazmi, 1989:40). The state played an active role at every level of economic development – from the construction of massive dams to provide power and feed irrigation schemes to the provision of subsidised industrial credit and fertiliser (Abrahamian, 1982). But overall, the Shah's economic policies did not bring substantial or lasting benefits to Iran.[18] The impressive output growth reflected primarily the exponential growth in oil revenues, much of which was wasted on military hardware, state bureaucracy, corruption and inefficient investment. The channelling of oil revenues into the productive sector via state support and subsidised lending to a select few was just as reckless as squandering money on military hardware (Nazmi, 1989). Misallocation of capital and the lack of diversification led the economy down the wrong development path, wasting scarce resources and undermining potentially more efficient producers. As a result, by the time the Shah was deposed, oil revenues accounted for 40% of national income and almost three-quarters of the government's revenue (Nazmi, 1989:51). Iran's non-oil income per capita was lower than that of many countries generally considered much poorer, "hardly appropriate for an economy the Shah named 'the new Japan' " (ibid.).

Attempts at economic reform

Rapidly accelerating inflationary pressures brought about by the oil boom prompted the authorities to engage in a disastrous programme of price controls and anti-profiteering measures which cost the Shah the support of the business community (Zonis, 1983:596),[19] and also stoked fears of sectarianism, as many of the businessmen accused of price fixing were members of minority communities (Jews and Baha'is) rather than Shiite Muslims. The anti-inflationary policies were also wholly ineffective, since the only long-term remedy to the accelerating inflation of the period – proper monetary discipline – was ignored. "The price controls . . . only led to shortages and the creation of black markets" (Nazmi, 1989:50).

As mentioned earlier, the Shah's key economic policy initiative, the White Revolution, launched in 1962 in response to the economic crisis of 1960 and pressure for liberalisation from Washington, had at its heart an extensive programme of land redistribution (Owen, 2000:92–3). This reform programme was symptomatic of the Shah's broad-brush approach to economic development. Motivated by a mixture of good intentions, hubris and an obsession with modernisation, and suffering from a total lack of attention to detail, the reform eventually alienated both the traditional landowners and the peasants who were supposed to be its main beneficiaries.[20]

These failed economic reforms, like the failed political reforms, only contributed further to the fall of the Shah in 1979.

The inequitable development path

The social tensions faced by the Shah in his attempt to reconcile the competing demands of the need for economic development, the drive to create broad-based development and the need to reconcile development with a socially conservative climate, are discussed in several studies of the Iranian revolution (Nasri, 1983). These tensions continued in the following decades. The Shah's authoritarian policies which focused almost exclusively on economic development, neglected – according to most observers – political reforms that would have promoted a more equitable development path and reconciled competing social and economic imperatives. Saikal, for instance, argues that the Shah's development programme lacked "a governmental system that would allow increasing political and economic decentralization, public participation, and individual initiative, and thus put more stress on social development and progressive redistribution of wealth" (Saikal, 1980:203–4).

Nasri points to the growing social dislocation caused by the rapid pace of economic development in the 1970s. However, he argues that it was this friction between economic and social/cultural imperatives (which affected all classes in Iranian society), rather than the disjunction between growth and broad-based development (which primarily affected the poor), that created the widespread hostility to the Shah's rule:

> It was . . . not the lack of emphasis on economic 'development' as against economic 'growth' that was the source of the problem . . . rather, it was the rapid and uncontrolled rate of development that not only made Iranians feel like aliens in their own country, but also increased their expectations to such levels that the government could not have satisfied them even under normal circumstances (Nasri, 1983:619).

Zonis (1983) cites evidence that the perceived inequity of the Shah's development programme was critical in creating opposition to his regime. He follows Walton (1986) in citing Hirschman's (1973) analysis of the effect of development and inequality on the political process. Walton argues that expectations of ever-rising material wealth were becoming increasingly frustrated and the people's tolerance for the continually rising social and economic inequities was becoming rapidly exhausted. Zonis takes the view that the "politicization of income inequality" was a key development of the second half of the 1970s, driven by a "panic over what were seen as disappearing opportunities for riches, and bitter envy directed at those who had already succeeded" (Zonis, 1983:597). Halliday (2003:50–53) shares this view, citing the significant increase in income inequality, which saw the urban poor in particular fare badly, as a key factor behind growing opposition to the Shah's regime.

The fall of the Shah

The dramatic collapse of the Shah's regime in 1979 appeared to catch most observers off-guard. Zonis notes: "The breathtaking speed of the final denouement caught not merely the Shah and the Americans off-guard; Ayatollah Khomeini himself had no idea of the speed with which the Pahlavi system would collapse in those final weeks" (Zonis, 1983:602).

For the purposes of this study, the historical events are relevant in as much as they inform our analysis on what the determinants are of Iranian policy regarding its oil and gas industry. In other words, what is the dynamic in Iranian political life and society that brought about revolutionary change? Additionally, what are the economic factors relevant to the revolution and vice-versa, and how would the revolution and the revolutionary discourse impact on economic policies?

Fred Halliday, one of the most reputed specialists on the Iranian revolution, identifies several key reasons that led to the Shah's overthrow (2003:50–67). Firstly, the rapid and uneven pattern of economic development increased inequality and left many bewildered at the rapid pace of change. People's expectations ran so high that the government could not satisfy them. Worse, the drop-off in economic performance at the end of the 1970s and the growing evidence of waste and corruption cemented peoples' impression that they had missed their chance for individual material progress. "When it became obvious in mid-1975 that the boom was slowing down, there was panic over what were seen as disappearing opportunities for riches, and bitter envy directed at those who had already succeeded" (Zonis, 1983:597).

It was a tension created by uneven development and an unfair one because of clientelism. Furthermore, there was the tension arising from the incongruity between authoritarianism and modernism. "The Shah was pursuing a capitalist-oriented process of socio-economic development [that] . . . unleashed forces opposite to those the throne required for its centrality in politics" (Saikal, 1980:203).

As early as 1960, a US government report on Iran, prepared for the newly elected President Kennedy, identified the Shah's core problem:

> Under the Shah, Iran has made considerable progress in economic development, in social welfare, and in internal security and administrative efficiency. The progress has, however, taken place without participation in the government by the main opposition groups. To some extent, the Shah's isolation from these groups has been due to his unwillingness to ride demagogic issues appealing to the lower popular passions . . . To an equal extent however, it has been due to his unwillingness to listen to critical advice, to his unwillingness to share power, and to his near-obsession with military affairs.[21]

These conclusions are just as true when discussing the fall of the Shah's regime in 1979.[22]

While there was considerable economic development under the Shah and there were even attempts at political and economic reform, the Shah's ability to rely on oil money and external backing allowed him to ignore domestic issues and dismiss the views of the opposition. The lack of involvement of other parties in government eventually led to the Shah's downfall. Halliday (2003) identifies the broad nature of the political coalition ranged against the Shah as a key cause of his overthrow. This point is reinforced by Katouzian, who argues that the 1979 revolution "shared the basic features of traditional Iranian revolts [in being] aimed at the overthrow of the state – indeed the person of the ruler – at all costs, [having] the support of the entire political society [and] not [being] resisted by any social class" (Katouzian, 2003:31).

To an extent, the breadth of the coalition illustrated the naivety of some of those involved – particularly secular and liberal forces, who underestimated the power of the clergy. Thus, despite a mismatch of objectives, the liberal and secular forces had hoped to steer the unfolding of the revolution towards their own objectives.

> Among the secular and middle-class forces many hoped that . . . they could deflect the movement away from its clerical patrons. This enabled such people to support the movement with appropriate optimism, but it represented an underestimation of the strength of the religious forces (Halliday, 2003:56).

The mobilising role of Islam and its personification in the charismatic persona of the Ayatollah Khomeini himself was also critical to the revolution's success. "Khomeini was in many respects the epitome of a charismatic leader" (Halliday, 2003:61). He was also a skilful political operator – able to play to different audiences and broaden his message – which explains in part why the broad anti-Shah coalition was maintained as long as it was.[23] He cites the ambivalent international context which allowed Khomeini to portray the Islamic Revolution as a nationalist movement in opposition to the West and the United States in particular (the Shah's patrons). The history of external interference – detailed in the previous section – helped to foster the "conspiracy mentality" which Khomeini played on (Halliday, 2003:63–7).[24] The American change in tone and direction under Carter's presidency – pressurising the Shah to liberalise when his regime was facing its greatest threat – was perhaps the final straw, particularly taken in tandem with the economic downturn which hindered the state's distribution of oil revenue to buy political favour.[25]

The Shah's fall underlined several important aspects of Iranian politics and social dynamics. Firstly, it showed that development does not necessarily create beneficial effects when not accompanied by democratic political reform. Clientelism and authoritarian rule impede fair distribution of wealth, as well as the implementation of consistent economic measures. Thus, while Mossadeq's nationalisation of the oil industry, for example, alienated some groups of the society, the subsequent relaxed economic agreements with respect to the oil industry did not create a more favourable environment. This was due to the fact that the administration of the country's revenue from oil was run by members of an elite who, when they did not use it for the growth of their own businesses, invested in military hardware and used it to consolidate their political power.

Other specific reforms, such as land reform, failed as a result of a combination of discretionary measures (e.g. to put farmers into collective farms against their will) and opposition from the conservative groups within society (landowners, clergy) (personal communication).[26] The conservative elements proved to be important in voicing the discontent with the regime. The fact that the revolution could rally considerable support through the use of nationalistic and anti-Western rhetoric is essential in understanding the shifts in the economic policies with respect to privatisation in Iran. It is also pertinent to the understanding of the type of political culture present at the time in Iran. It actually points out that Iran was at that stage a largely "traditional society", where politics was dependent on the skills and charisma of the ruler/leader rather than on the effectiveness of the institutional framework. However, one must also take into account the fact that the revolution fundamentally changed the political and economic functioning in Iranian society and therefore the determinants of policy, including the policy concerning privatisation. The next chapter addresses the changes in Iranian politics and economics.

To sum up, the issues discussed in this chapter are essential to the understanding of the interactions between the political, economic and social structures in Iran. It has identified an authoritarian rule based on discretionary measures as one of the features of Iranian political life.

Clientelism and uneven development ensued, creating an ever-increasing gap between the rich and the poor.

The troubled relationship with Western players, as well as the temptation to find a scapegoat for the structural economic problems, nurtured an anti-Western feeling and a strident nationalistic rhetoric in society, a circumstance that would have serious repercussions on post-revolutionary policies.

Oil, the cornerstone of the Iranian economy, was used not only by the Shah but also by the revolutionary regime. By centralising power, the Shah became the only individual who could make decisions concerning the privatisation of oil and gas. Control of oil profits allowed the Shah to also control the economy and bolster his political power at home. His ambition to raise his stature on the world stage led him to overspend on military hardware to the detriment of the economy. Unfortunately for him, a subsequent drop in oil prices undermined economic development at home and so deepened discontent among the people, creating a fertile ground for revolution.

NOTES

1 Reza Shah was forced by Britain and the Soviet Union to abdicate in favour of his son, Mohammad Reza Shah, in 1941.
2 A British government source from the same period claimed that "no two Persians can ever work together for any length of time, even if it is jointly to extract money from the third party" (Abrahamian, 1982). Ignoring the patronising Foreign Office hauteur and orientalist tone, this may still be an accurate portrait of Iranian political life at the time, not because of some endemic fault in the national character, but because of legitimate and rational ideological divisions.
3 For example, some twelve Prime Ministers were installed in the course of 13 years.
4 According to the *Encyclopaedia of the Orient*, ulama is used for "a group of men with religious education and religiously related professions. Ulama is the group of men expressing the true content of Islam towards both the people and the rulers" (*Encyclopaedia of the Orient*, available on-line at www.Lexicorient.com). See also http://i-cias.com/cgi-bin/eo-direct.pl?ulama.htm.
5 Personal communication. Discussions with Mr Hossein Kazempour Ardebili, member of the Board of Governors, OPEC, and Head of the Iran delegation to OPEC, December 2005.
6 Ibid.
7 Katouzian goes on to note, however, that in most cases substantial involvement by Iranian forces or miscalculations on the part of the Iranians played a significant

role. Nevertheless, even the Shah eventually came to believe that he was the victim of a malign foreign (CIA) plot: "the Shah . . . fell back on the old Iranian formula of seeing Iranian politics as some kind of vibration elicited by foreign players" (Zonis, 1983).

8 Personal communication. Discussions with Seyed Mohammad Hossein Adeli, former Iranian Ambassador to the UK and present editor of *Economic Trends*, Tehran, February 2006.

9 As Zonis argues, the Shah even believed that the CIA was behind the emerging unrest in the 1970s, falling back "on the old Iranian formula of seeing Iranian politics as some kind of vibration elicited by foreign powers" (Zonis, 1983:595).

10 He was also supported by the Soviet Union and China – indeed, the Shah's last state visit was to Beijing (Zonis, 1983). However, it was his relationship with the West which had the greatest effect on society.

11 Personal communication. Discussions with Seyed Mohammad Hossein Adeli, former Iranian Ambassador to the UK and present editor of *Economic Trends*, Tehran, February 2006.

12 Abrahamian (1991) quotes a number of Khomeini's anti-Western slogans from the time of the revolution. The dictum, "the oppressed nations of the world should unite against their imperialist oppressors" gives a flavour of their tone and content.

13 Personal communication. Discussions with Seyed Mohammad Hossein Adeli, former Iranian Ambassador to the UK and present editor of *Economic Trends*, Tehran, February 2006.

14 Personal communication. Discussions with Seyed Mohammad Hossein Adeli, former Iranian Ambassador to the UK and present editor of *Economic Trends*, Tehran, February 2006.

15 There were specific changes in the economy during the Shah's rule which reinforced this autonomy and supremacy of the state combined with a capitalist class. In the aftermath of the 1953 coup, the Shah abandoned import substitution policies, shifting to a less protectionist policy that generated a substantial trade deficit and hit domestic producers hard (Nazmi, 1989). As domestic producers withered in the face of fierce competition, state investment increasingly substituted for private sector development, rather than aiding it. Following a severe recession and balance of payments crisis in 1960, import restrictions were imposed as part of an IMF-led austerity programme (ibid.). The government augmented the IMF-imposed measures with restrictions of its own that essentially saw Iran returning to the classic import substitution path to development. As ever, it was financed through the economic rents earned from oil.

16 Ironically, the Shah's military build-up and forays into the foreign policy world "estranged many foreign leaders" (Nasri, 1983:624).

17 Arms dealers "began to jest that the Shah read their manuals in much the same way as other men read *Playboy*" (Abrahamian, 1982:436).

18 However, on a more serious note Abrahamian argues that, although "substantial sums were squandered on palaces, royal extravagances, bureaucratic consumption, outright corruption, nuclear installations and ultrasophisticated weapons too expensive even for many NATO countries" (Abrahamian, 1982:427), much of the oil wealth was channelled into more productive uses.

19 Some 8,000 businessmen served prison sentences of up to three years, and 250,000 fines were issued (Abrahamian, 1982:498).

20 As Owen observes, the reform tried to force peasants into collective farms, creating a "legacy of bitterness that . . . meant that the regime lost all possibility of obtaining political support from the direct beneficiaries of the reform" (Owen, 2000:93). The White Revolution also alienated the clergy, who were significant landholders (Nazmi, 1989). This group also objected to the extension of women's rights and literacy projects that were included in the reform programme.

21 Quoted in Zonis (1983:587) from a report by John W. Bowling, Deputy Director of the Office of Greek, Turkish and Iranian Affairs.

22 In fact, most commentators on the Shah have echoed these comments, even up to the present day. Milani (1997), for instance, argues that the Shah "wished to be the king of a modern country, but he was determined to rule using a medieval system of governance. That is why he modernised the economy but failed to do the same for the political realm."

23 Khomeini succeeded in mastering populist rhetoric that helped to soften his Islamist message. As Owen notes, he was "a master of a type of populist rhetoric, with its emphasis on such central themes as anti-imperialism, democracy and social justice, that seemed to provide a consensus around which all opponents of the Shah could come together" (Owen, 2000:94). This point is made more forcefully by Abrahamian (1991:106), who argues that "we should recognise Khomeini-ism as a specifically religious variant of populism".

24 Abrahamian quotes a number of Khomeini's anti-Western slogans from the time of the revolution. The dictum, "the oppressed nations of the world should unite against their imperialist oppressors", gives a flavour to their tone and content (Abrahamian, 1991:114).

25 Nasri (1983) comments on this issue in more depth, quoting a number of authors in arguing that Carter's policy towards Iran confused the Shah, held up the sale of riot control equipment and gave succour to the opposition.

26 Personal communication. Discussion with Dr Manouchelir Lighrani, M.D., former landowner, Tehran, December 2004.

3

REVOLUTION AND REGRESSION, 1979–89

Introduction

The discussion so far has shown that there were several relevant issues that needed to be considered in order to understand the interactions between the political, economic and social structures in Iran. These include the existence of an authoritarian regime, sustained by repression and clientelism, the regime's inadequate interaction with society at large and the maldistribution of wealth consequent on clientelism

The focus on economic modernisation without radical political reform alienated different sections of Iranian society. Even well-intentioned schemes, such as land reform, failed because of arbitary measures being imposed on those it intended to benefit, and also because of opposition from conservative forces. The conservative elements (clergy and landowners) proved to be effective in voicing discontent with the regime. They also rallied against the regime's societal groups (such as students) which had quite different priorities. The unifying element was the nationalistic and anti-American sentiment, and a deep tradition of contesting those in power. The fact that society responded to the charismatic figure of the Ayatollah Khomeini points to the existence of a largely traditional society, where political inclinations depend on the personal qualities of the leader rather than on the effectiveness of national institutions.

As for the oil industry, it must be said that it had been largely politicised. Oil was used as a political tool by both the Shah and the revolutionary regime. Being the main source of revenue, it was obvious that its control by the leadership would enhance political control at home and, being a strategic commodity, could boost its status abroad.

This chapter analyses the post-revolution period between 1979 and 1989, along the same coordinates as the previous chapter. During this period, we see the major post-revolutionary changes in Iran, specifically in its economic system, taking shape. It seeks to answer the question of why the oil and gas industry was renationalised after the revolution and how the type of privatisation evolved and changed over time. It focuses on the political forces that determined economic policy and vice versa in order to answer this question. It details the nature of political change in Iran. In addition, Khomeini's political views are examined in some detail as well as those of competing political constituencies. These factors, combined with a look at the major foreign policy issues faced by Iran, give the contextual background to the country's economic decisionmaking. The second section of this chapter examines the economic policies of the post-revolution government and developments within the economic system.

The Iranian revolution was not just a political revolution, it was an economic one as well. The revolution heralded an entirely new relationship between society and government. In particular, the clergy adopted a much larger social role. Khomeini's politics walked the line between fundamentalism and populism. This was also true of the economic direction, where Islamic economic ideals were mixed with populist ideas about the redistribution of wealth. While the concept of private property was never really questioned, much of Iran's industry was nationalised after the revolution.

Studies of the early post-revolutionary period focus on the creation of the Islamic Republic under Khomeini's charismatic leadership. They analyse the role of the various political forces allied, sometimes loosely, to Khomeini's leadership and outline the emergence of the key ideological fractures within the new political establishment, divisions that remain to this day. While seemingly obvious, we must ask to what degree the Iranian revolution really changed the political system in Iran. Clearly, those in high political positions changed and the influence of the clergy became much more pronounced. But did the revolution change the relationship between government and society, and so with the economy, or was the change merely a question of different institutional rhetoric?

A Novel Kind of Revolution

The reality of the Iranian revolution contrasted markedly with the conception of revolution that had prevailed since the late eighteenth century. It rejected progress, material improvement, national assertion, historical legitimacy and democratic sovereignty – five themes which, however violated in practice, were at least invoked formally by modern revolutions from 1789 onward (Halliday, 2003:45). It was not a revolution "in the proper sense of the word; what happened in Iran was a comprehensively *reactionary* revolution, restoring to the term its original, astronomical, meaning of *return* to a previous order" (Halliday, 2003:43).

The revolution in Iran during 1979 shared many of the features of other Third World uprisings. It drew support from sections of the civil service and trading communities, and from the poor urban population. It mobilised against a dictatorial regime and centred around a charismatic leader and an ideology of revolutionary legitimacy.

The brevity of the revolution and the continuity of thought

Unlike many other revolutions that took years to complete in full, the Iranian revolution happened at high speed. Within less than a year and a half, it leapt from protests to the declaration of the Islamic Republic. The timeline of the changes in the structure of the regime suggests that all the contradictions remained to be worked out after the revolution. In fact, many can still be traced to the present. Briefly these are:

- January 1978: street protests begin. They are organised by religious students in Qom who are protesting against an anti-Khomeini newspaper article.
- Strikes and protests in Iranian cities occur over the next few months. They are organised by the clergy and supported by the bazaars.
- September 1978: the religious processions after the fasting month of Ramadan swell to become large political demonstrations.
- 8 September 1978: martial law is imposed and the demonstrators are shot by the security forces.
- October 1978: the wave of strikes begins. The strikes spread to become a general strike, in late November 1978. Significantly, the oil fields, which are at the heart of the Iranian economy, are shut down.
- 5 November 1978: the Shah is under pressure from the military. Eventually he appoints a military government which fails to end the strikes. In December, the government is forced to allow street demonstrations to take place. This happens during the Shi'ite festival of Ashura.
- Throughout this period, Khomeini leads the protests from his exile in Paris.
- 15 January 1979: the Shah leaves Iran. The former opposition leader and secularist, Shahpour Bakhtiar, becomes head of the government.

- 1 February 1979: Khomeini returns to Iran from Paris. He refuses to negotiate with the Bakhtiar government. Khomeini appoints Mehdi Bazargan, a liberal Islamist, as head of a rival provisional government.
- 10/11 February 1979: pro-Khomeini mutinies break out in the armed forces. The army declares itself neutral. The Bakhtiar government collapses.
- 30 March 1979: the Islamic Republic is declared following a referendum.
- November 1979: an Islamic constitution is passed by referendum. Khomeini is officially accepted "supreme leader", not only *faqih* but *valy-e faqih*.
- January 1980: Abol-Hasan Bani-Sadr, an Islamic socialist, is elected President. Following its election the first Majlis (which is dominated by the Islamic Republican Party) selects Mohammad Ali Rajai as Prime Minister.

The short span of the revolution did not leave enough time for Iranian society to evolve from being ruled by the secular Shah to being governed by a theocracy. This meant that many of the same factors determining politics and economics in Iran continued into the post-revolution period (personal communication).[1] These included the same factors discussed in the previous chapter, including arbitrary rule, perception of the West, and the relationship between economic and political development. However, the combination of these determinants of policy and other specific new factors make them appear different. At the same time, religion and revolutionary politics played increasingly important roles as determinants of policy.

The New Political System: Continuity and Change

As mentioned earlier, the main elements of change introduced to the system of government were to reflect a blend between the principle of popular sovereignty and theocratic rule. Thus, the 1979 constitution was an amalgam of the 1906 Iranian constitution and that of the French Fifth Republic, with further amendments to incorporate Khomeini's interpretation of Islamic government.

However, these changes created contradictions and ambiguities. As a number of commentators have noted, the constitution represented an uneasy compromise between secularist notions, such as the separation of powers and Islamic principles.[2] For instance, the constitution created a (relatively democratic) parliament with roles for a Prime Minister and President. However, these democratic and essentially secular features were counteracted by the creation of the Council of Guardians – a twelve-member body composed of clerics and Islamic jurists with veto power over the Majlis. This power of veto would provide the key point of conflict between radicals and conservatives within the fundamentalist Islamist grouping once its grip on power was assured, in 1981. The other essentially non-democratic feature of Iran's 1979 Islamic constitution was the creation of the position of *faqih* in line with Khomeini's notion of *velayat-e faqih* – a post which Khomeini himself held until his death in 1989.

The post-1979 political system can be characterised as "limited popular sovereignty" (Milani, 1997:82), in that it had significant democratic elements (particularly in relation to the regional and historical norm), but this nascent popular sovereignty was "disfigured" by five key constitutional features:

- The constitution put God, not the people, as the ultimate authority;
- it provided a hierarchy of power, with God at the top, the people at the bottom, and, crucially, the clerical establishment in the middle;
- it granted unelected bodies, namely the Council of Guardians and the *faqih* powers of veto and the right to appoint officials over democratically elected politicians;
- it limited the power of the President (the only nationally elected figure) by granting the *faqih* powers to sack him and to command the armed forces;
- it curtailed freedoms of the press and of assembly if deemed inimical to Islam.

This "limited popular sovereignty" is in line with traditional Shi'a ideology which holds that only the highest clergy can deduce the Qur'an's secret meaning (*baten*) from its apparent meaning (*zaher*).[3]

The above-mentioned blend generated not only a specific ideological make-up in Iran, but also increased factionalism within the government, and created political deadlocks between the radicals and the conservatives. The creation of parallel institutions signalled that the structural features of the previous regime (repression of dissonant voices) still persisted, despite the difference in the political institutions. However, unlike the pre-revolutionary period, now the Islamic and populist/socialist rhetoric was deeply felt and impacted upon the Iranian economic decisionmaking process (personal communication).[4]

Ideological make-up

The ideological make-up of Iran reflects the mixed tendencies to create a state based on popular sovereignty and theocratic institutions. Due to its ideological diversity, Shi'ism spawned variations in the political thought of various groups. Thus, the post-revolutionary clerical establishment (including Khomeini himself) was amenable to borrowing elements from the various ideological subsystems within Shi'ism so as to enhance its position of power, rather than adhere to a uniform system. This does not mean that members of these Shi'a groups can be classified according to their proclivities. As Milani puts it:

> Failure to grasp the flexibility which the adoption of different aspects of various subsystems offers to the clerical leadership leads to erroneous categorisations such as Islamic modernists, traditionalists, fundamentalists, moderates, extremists, etc. (Milani, 1997:82).

Thus, the clerical leadership can be seen as having a range of tools to draw on. In particular, there were four distinct ideological strands within Iranian Shi'ism that have informed the establishment's governing programme since 1979. The first strand, linked to Ayatollah Morteza Motahhari, is an essentially conservative interpretation of Shi'ism that rejects the Marxist elements incorporated into Islam by the *Maktabis*. Motahhari was a leading cleric, a member of the influential Revolutionary Council and a close confidant of Khomeini.[5] His key motivation was to counter the anti-clerical currents common among young Islamic revolutionaries; he also had a "deep respect for the market system based on private property . . . [which] removed the anxieties of property-owning classes" (Rahnema and Nomani, 1990:42). This defence of private property placed him in the *Hojatieh* tradition, although his emphasis on the political role of the clergy – embodied in Khomeini's notion of *velayat-e faqih* – was not universally popular with conservatives (Ehteshami, 1995:9).

Rahnema and Nomani outline a second strand of Iranian Shi'ism at the time, which they link with the work of the intellectual Ali Shariati, a nationalist supporter of Mossadeq in the 1950s. Shariati's Shi'ism borrowed extensively from Marxism–Leninism, an ideology the pro-Moscow Tudeh party had embraced. The authors go as far as to argue that "Shariati's achievement was the utilisation of Islam as a socially applicable . . . form of presenting a content which could be considered as essentially Leninist" (Rahnema and Nomani, 1990:53). Indeed, Shariati emphasised the discordance between the oppressors and the oppressed, a notion that formed an essential part of Khomeini's own pre-revolutionary rhetoric (Abrahamian, 1991). However, Shariati differed from Khomeini in that he advocated the establishment of an Islamic state but without clerical leadership (Rahnmena and Nomani, 1990:62). The Islamic Republic's leadership arguably borrowed from Shariati's rhetoric, particularly in building support for the revolution during the final years of the Shah. Shariati's belief in a "totalitarian transitional government" – derived from an "elitist and paternalistic position in relation to the people" – can also be interpreted as "very similar to Khomeini's concept of an Islamic government" (Rahnmena and Nomani, 1990:67). However, Khomeini was also careful to uphold property rights in order to avoid alienating his conservative business constituency.

A third, essentially conservative reading of Shi'ism is embodied in the career of Sayyed Mujtaba Mir-lowhi (known as Navab-Safavi), the founder of the radical Islamic group Fadaian-e Islam that violently opposed the Shah. The authors added: "Navab-Safavi's subsystem represents a violent and reactionary Islam which contains many of the necessary attributes of a totalitarian system" (Rahnmena and Nomani, 1990:78).

This reactionary reading of Islam – a demand for the strict implementation of the *sharia* under a theocratic government – proved popular amongst the huge numbers of rural migrants to Tehran and other cities, the groups most alienated by the rapid cultural and economic change instigated by the Shah's development policies. Migrant workers reacted to this "alien cultural environment" by forming *hey'ats* (informal religious groups), which went on to form important power bases for the Islamic Republic's rising political star. This "vast source of bigoted militancy" would become the "vengeful executive arm of the clergy" (ibid.:88), referring to the *Komites* (vigilante police) and *Pasdaran* (revolutionary guards) who participated so effectively in the post-revolutionary terror (Milani, 1997:84).

Finally, there is a more liberal reading of Islam, associated with the veteran nationalist, Mehdi Bazargan, the first Prime Minister under the Islamic Republic. Bazargan tried to blend Islam with liberal democracy. Essentially, he argued for a constitutional theocracy in which religious observance was a matter of individual conscience with the clergy having little or no political role (Rahnema and Nomani, 1990:99–116). Although Khomeini had links with Bazargan – as evidenced by the latter's appointment as Prime Minister in February 1979 – little of Bazargan's liberalism found its way into the Islamic Republic's political direction after the more conservative clerical establishment had tightened its grip on power in subsequent years. However, elements of Bazargan's more liberal reading of Shi'ism can arguably be identified in the philosophy of President Mohammad Khatami (elected President in 1997), and the views of his brother and key advisor, Reza Khatami. Against this background, therefore, it becomes all the more important to rethink the concept of development and perceive development in the context of a particular society with its culture, its religious disposition, values and, in the case of Iran, its rapidly changing social fabric.

Parallel institutions

The specificity of the institutions created by the revolutionary regime does not necessarily imply a break with the practices of the previous regime. On the contrary, the emergence of parallel institutions under the leadership of Khomeini and the Islamic Republican Party (dubbed Khomeini's "mini-state") (Milani, 1997) was motivated by the same drive to consolidate and preserve power. Repression was the means to achieve it. Khomeini's mini-state drew much of its support from the culturally alienated rural–urban migrants – the "vast source of bigoted militancy" (Rahnema and Nomani, 1990:88) – and was funded by Khomeini via the Mostazefan (Oppressed) Foundation, which had expropriated the assets of the royal family and their associates. The mini-state adopted a range of unofficial organisations, including the *Komites*, the *Pasdaran* and the Revolutionary Courts, which dispensed brutal summary justice to those deemed opposed to the pro-Khomeini forces (Milani, 1997:84–5).

The mini-state had effectively neutered all opposition by 1983, moving against the secularists and Islamic Liberals associated with Mehdi

Bazargan's interim government, and then, in 1981, forcefully ejecting the Islamic Socialist Bani-Sadr, who had been chosen by Khomeini himself in 1980 as the Islamic Republic's first President, from office. Even senior clerics were not immune. Ayatollah Seeyed Kazem Shariatmadari, who had opposed the *velayat-e faqih* provision of the constitution, was "virtually put under house arrest until his death in 1986" (Milani, 1997:85). After the dismissal of Bani-Sadr, the clergy's control over the Islamic Republican Party, and therefore the government, was almost total (Rahnema and Nomani, 1990:255). This ushered in a period of "pragmatic clerical rule", with elements close to the spiritual leader forcing the more militant members of the clerical coalition to tone down their revolutionary rhetoric and bring an end to the "semi-anarchy" of the "mini-state's parallel institutions" (ibid.:256).

Factionalism and its effects on political and economic policies

Factionalism did not act as a determinant of policymaking in Iran. It was related to Khomeini's own ambiguous political views, which drew on different trends within Shi'ism. The swing between popular sovereignty and theocratic rule represented an additional factor (Milani, 1997). During the revolutionary period (1977–9) and the early years of the Islamic Republic, political participation was at its peak. The revolutionary movement itself was initially extremely wide-ranging, with factionalism temporarily in abeyance. As Katouzian argues, the Islamic Revolution (in common with the 1905–9 Constitutional Revolution) had "the support of the entire political society" (Katouzian, 2003:31). Milani characterises the 1977–9 period as "the golden age of Iran's mass politics" (Milani, 1997:83). However, he goes on to catalogue how Khomeini created a political mini-state within the broad coalition of anti-Shah forces, which then went on to destroy its revolutionary comrades one by one.[6] With political participation outside the establishment outlawed, it took "the form of involvement in factional politics" (Milani, 1997:85).[7]

Factionalism within the Islamic Republican Party resulted in part from the dominant role in government of Khomeini himself – as *faqih*. As Milani (1997) observes, Khomeini kept the party's various factions competing with each other in order to maintain his own position at the top of the republic's power structure. Khomeini eventually tired of the infighting, which had resulted in a damaging legislative gridlock, and disbanded the Islamic Republican Party. However, the demise of the party did not bring an end to factionalism, which continued to hamper the implementation of the radical legislative agenda of the left-leaning faction that considered itself closest to Khomeini. To this extent, factionalism under Khomeini reflected the fact that, despite being the Islamic Republic's figurehead and most gifted political operator, he was unable to maintain total control over the regime (Owen, 2000:181). In particular, he remained constrained by powerful conservative elements within the clerical establishment who were opposed to the more radical "Islamic Socialist" proposals that initially attracted so many Iranians to the Islamic Republic's banner (Milani, 1997:87).[8]

There was also a deeper issue that contributed to factionalism, and that was Shi'ism. One can say that it did not only provide a variety of tendencies within Iranian society with respect to different issues (the role of the state, property, etc.) but also determined or envisioned the type of political action in line with Khomeini's vision. It nurtured, in other words, a variety of potential policies on private property, for example, but also accounted for the failure of the system to function by fostering factionalism and causing political deadlocks. Shi'ism has traditionally been a "polycephalic faith" (Milani, 1997:86). Shi'ism, unlike some other religious orders, and to a degree unlike Sunniism, has neither a church nor an accepted power hierarchy. The Islamic constitution of Iran originally stipulated that the *faqih* must be a *marjaetaqlid* and also be accepted by the majority of the people. This was an attempt to create a hierarchy of religious as well as political powers – something which had never existed before. But even under Khomeini, other grand ayatollahs (Seyyed Kazem Shariatmadari and Mohammad Reza Golypayegani) had large mass followings. While there is no need to cover the revision of the constitution in 1989, it also played a role. The revision meant that the *faqih* was no longer required to be a *marjaetaqlid*, nor be accepted by the majority of the people as such. Hence, in 1989, the Assembly of Experts was able to select Seeyed Ali Khamenei as the new *faqih*, even though *hojatolislam* was ranked below *ayatollah*. At the same time, Khamenei had declared that he was not seeking religious leadership in Iran. As Milani notes,

> this fact is testimony to the Islamic Republic's failure to render Shi'ism monolithic, as there are many people who follow not Khamenei but other ayatollahs in Iran and Iraq. Therefore, factions and Shiism are inseparable, like two sides of the same coin (Milani, 1997:86).

Grounded in the "diversity of [the Islamic Revolutionaries'] socio-economic interpretations and subsequently programmes", as well as chronic factionalisation characterised the first decade of the Islamic Republic (Rahnema and Nomani, 1990:37). During the immediate post-revolutionary period, three distinct factions of "republican contenders" emerged and competed intensely for power. Of these groups, the Fundamentalist Islamic Republicans and Liberal Islamic Republicans quickly neutralised the Secular Republicans. By 1981, the Fundamentalist Islamic Republicans had also successfully eclipsed the Liberal Islamic Republicans (Ehteshami, 1995:8).

The Fundamentalist Islamic Republican group itself consisted of several competing factions, ranging from the *Maktabis* (radicals who combined elements of Marxism with traditional Shi'ism) to conservatives centred on the *Hojatieh* society (more traditionalist clerics allied to business interests). Between these competing extremes were more pragmatic, career-oriented politicians, many of whom rose to prominence following Khomeini's death in 1989 (personal communication).[9]

The explosion of factionalism within the ruling coalition made policy implementation (particularly with regard to economic policy) virtually impossible.[10] As discussed above, the critical stalemate was over the issue of private property, where radical groups within the Majlis saw their attempts to introduce anti-capitalist policies consistently blocked by the conservatives in the Council of Guardians. The policy log-jam period was described by Rahnema and Behdad as follows:

> The arena of public property [whereby] the strict interpretation of Islam, in defence of the propertied class and in support of the conservative political forces, gains the upper hand over any reading of Islam in support of the populist political forces. This confrontation is seen in the experience of the Islamic Republic of Iran, where from the summer of 1980, when the Majles was established, to the end of the third Majles in the summer of 1992, all the main reform laws passed by the populist parliament, such as land redistribution, nationalisation of foreign trade, and the labour law, were rejected by the Council of Guardians as being incompatible with Islamic precepts (Rahnema and Behdad, 1995:12–13).

The 1987 labour law was "the point of greatest intensity in the economic dispute between the two factions" within the clerical establishment (Rahnema and Nonani, 1990:267). The Council of Guardians rejected half of the 200 articles of the proposed law as un-Islamic. The Council continued to obstruct the law's implementation even after interventions by Khomeini in its support. The divisions over land reform were intense, and "such was the strength of conservative feeling that a bill proposing only a moderate act of rural redistribution . . . was held up as un-Islamic by the Council of Guardians from 1980 onwards" (Owen, 2000:95). The cause of this institutional gridlock was the sharp division between two competing interpretations of Islam in the area of property rights, and the existence of competing power-bases controlled by these different ideological factions embodied in the Islamic Republic's 1979 constitution. Thus, the implementation of a stable, consistent programme of economic reform under Khomeini was prevented by political conflicts "and the existence of semi-anarchic decisionmaking centres" (Rahnema and Nomani, 1990:269).

Features of the Iranian State

As noted in the previous chapter, the arbitrary rule of the Shah characterised Iran's inability to forge a stable political system (Katouzian, 2000; 2003). In other words, arbitrary rule signified a disjunction between the state and society in the country; rather than representing the interests of one particular class or group, the state exercised arbitrary power over all groups in that society, outside the confines of an established set of laws (Katouzian, 2003). The use of force and discretionary measures were necessary mechanisms for arbitrary rule. Clientelism was a direct consequence of such a state, as it allowed benefits to go to only a small group of followers in exchange for their support of the regime. It also conditioned the business relations with other actors, since the state was not an equal partner, but one which had the power to impose and change the rules of the game while still playing.

An interesting element of the post-revolutionary period, on the other hand, was the attempt to combine popular sovereignty with clerical rule. It will be shown that the constitution of the Islamic Republic had, at its core, a contradiction between popular sovereignty and the role of the clerics. The state's theocratic foundations were laid when the constitution endorsed the concept of *velayat-e faqih* and the *faqih* and institutionalised the Council of Guardians (Rahnema and Behdad, 1995:11). Thus, although the Islamic Republic provided for limited popular sovereignty – an opportunity to change the extant relationship between the state and society – the constitution conferred primacy of power on the clerics over democratic institutions, stating that the framers "created a system that is more Islamic than republican" (Milani, 1997:82).[11]

The existing constitutional arrangement, although no longer underpinning an authoritarian state, did share some notable features with one. It is true that the clerical establishment is not wholly removed from wider society, since significant factions within it command support from various parts of society. Nonetheless it wields *almost* arbitrary power and this has the potential to alienate large swathes of the population in the manner of previous Iranian rulers. Despite the persistence of these features of the Iranian state, one may also say that the revolution did make some fundamental changes in the relationship between society and government. The next question to ask is: what were the new factors which affected the government's decisionmaking process?

The post-revolutionary government

The post-revolutionary government needs to be understood along two lines. Firstly, there was a fundamental tension between the secular nature of the administration and its religious basis. Secondly, the revolution changed the political landscape and brought out new constituents which influenced policy. By analysing the politics of Iran with these considerations in mind, it is possible to discern the effect that politics had on the evolution of oil policies. In particular, how Islam, or rather the Iranian understanding of Islam, had a bearing on the nationalisation of oil, how the various constituencies within the Iranian government created the post-revolutionary economic system, and how they contributed to its change over time. The tension between politics and religion was partially attributable to the ambiguity of Khomeini's political views which may be described as falling between a fundamentalist form of rule and a populist one. While Khomeini was not able to overcome this tension, he did understand it (personal communication).[12]

Khomeini's political views: populism, Shi'ism and popular sovereignty

The role of the leaders in decisionmaking processes (with respect to both foreign and domestic policies) has long been highlighted in the literature on foreign policy analysis (Jervis *et al.*, 1985; Anon, 1997). This is especially so in more traditional states where authority is centralised, the political system not very diversified and allegiances based on belonging to the same community and sharing a common set of beliefs; in such a setting, strong and charismatic leaders are expected to fully understand the overall interactions of politics and economics. Ayatollah Khomeini was both strong and charismatic. One should remember that this man was nominally in a position of leadership (as *faqih* and as the revolution's undisputed figurehead), who would unite the political factions, but his own philosophical outlook was difficult to interpret and his loyalty seemed to switch from one faction to another over time. This was reflected in a rather unstable political life, with the criteria of "legitimate" rule changing on the whim of one man. Unlike the Shah, Khomeini appears to have appreciated the tension between economic development, on the one hand, and culture and equity, on the other (personal communication).[13] However, for all his formidable political skills and initial mass appeal, he failed to find the means by which to overcome the tensions between the worlds of the material, the political and the cultural or spiritual.

Khomeini certainly succeeded in giving a political voice to the cultural and political tensions raised by the Shah's development policies. As a noted cleric and long-standing opponent of the Shah's policies, Khomeini personified the cultural opposition to the Shah's programme of economic development. Zonis (1983:591) notes that "Khomeini was widely believed to embody many of the central values of Iranian culture, values that the Shah was seen as having desecrated". He was also, perhaps uniquely, able to create a unified political message that brought together a traditional Islamist critique of Western-style capitalism with a radical rhetoric that was, if not Marxist, then certainly populist. "We should recognise Khomeinism as a specifically religious variant of populism" (Abrahamian, 1991:106).

This mix can be seen to have produced tension between fundamentalism and populism. Fundamentalism, which originated in a description of early twentieth-century American Protestantism, refers to a conservative orthodoxy concerning religious issues. Populism in the Iranian context, and others, refers to an ideology calling for the mobilisation of "the lower classes, especially the urban poor, with radical rhetoric against imperialism, foreign capitalism and the political establishment" (ibid.). Its proponents are predominantly members of the middle class.

Iran is generally seen as a fundamentalist state, though, as gauged earlier, Khomeini's political views occupied the middle ground – between populist and fundamentalist tenets, allowing him to move from one position to another.

Unlike Sunniism, Shi'ism does not present a consistent theory of the state. The Sunnis have traditionally associated religious duty with obedience to the state. Shi'ism, on the other hand, is divided. Some believe that the existing states should be seen as usurpers (of the Hidden Imam) and therefore should be shunned, while others acknowledge the state in varying degrees, from grudging acceptance – in the sense that a state is better than no state – to a complete acceptance of the legitimacy of the state similar to the Sunni stance.

Opinion appears divided as to Khomeini's views (personal communication).[14] His sympathies may have lain more with the radical Maktabi tendency of the revolution. Correspondence between Khomeini and Khamenei in 1988 over the role of the state shows that Khamenei, widely viewed as a conservative champion among the clergy, had argued for a limitation to the state's role on Islamic grounds. Khomeini rejected the notion, countering that the Islamic Republic's government was "a supreme vice-regency bestowed by God", one that could interpret Islam and so should take precedence over "all peripheral divine orders" (Halliday, 2003:69). His earlier writings reflected vacillation and vagueness. By the 1960s, his views had turned definitely against the monarchy. He began to argue that only the *ulama* and *fuqaha* (religious jurists specialising in *sharia* law) had a legitimate right to rule. This stance naturally led to the likelihood for populist opinion to take shape and made revolution an acceptable option.

His views on society began to change. Initially he believed in the more traditional notions of society, which agreed with the principle of individual ownership of property and a God-given hierarchy within society that should be respected, though not flaunted. After the 1970s he took a more egalitarian view, distinguishing the oppressors (rich) from the oppressed (poor) (Abrahamian, 1991). His new social views were then applied to the Shah, whom he accused of widening this divide. These accusations relied on populist rhetoric and can be seen as a switch from a more traditional, conservative approach to one which more or less explicitly relied on populism. During the 1978–9 build-up to the revolution, the political slogans of Khomeini and his allies increasingly resembled the kind of rhetoric more closely associated with leftist politics: attacking the widening inequalities under the Shah, calling for an end to oppression and class differences and broadening the message of the revolution to embrace not only Muslims, but the "oppressed nations of the world" generally (Abrahamian, 1991:112–15).

However, as Abrahamian notes, once in office Khomeini was unable, or unwilling, to implement the kind of radical economic reform that initially attracted so many leftists (Islamic and secular) to his cause. This could have been predicted as, even at its height, Khomeini's populist rhetoric was "extremely vague on specifics and silent on the question of private property" (ibid.:113). Or when Khomeini's pronouncements on private property were uttered, they were far from coherent. The same Khomeini who stated that "Islam will eliminate class differences" and "we are for Islam, not for capitalism" (ibid.:114) also argued that "as long as there is Islam there will be free enterprise" (Ehteshami, 1995:8). At the same time, in the views of at least one analyst, Khomeini's overall goal was "to invert Islamic authority so as to free the actions of the state from any Islamic restrictions, particularly with regard to property" (Halliday, 2003:70). In this regard Abrahamian writes:

> Despite the radical rhetoric, the [Islamic Republic's] Constitution undertook to safeguard private property . . . What is more, it intentionally avoided the socialistic phrase *nezam-e towhidi* (unitary order), a term which the Mojahedin and other Islamic radicals – as opposed to populists – wanted enshrined into the republic's Constitution (Abrahamian, 1991:117).

Abrahamian goes on to describe Khomeinism:

> Khomeinism – like Latin American populism – is mainly a middle-class movement that mobilises the masses with radical-sounding rhetoric against the external powers and entrenched classes. But in attacking the establishment, it is careful to respect private property and avoid concrete proposals that would undermine the petty bourgeoisie (Abrahamian, 1991:18).

These contradictory views on the role of the state and private property may be explained by reference to Khomeini's mix of Latin American populism and traditional Islamic and socialist ideas. However, Khomeini drew on these different discourses instrumentally. One can safely say that his principal aim was to gain and maintain political power for the clergy.

The leftist rhetoric in the run up to the revolution was therefore intended to garner support from the Left. "Religious fundamentalism would never have won such support" (Abrahamian, 1991:118).

Similarly, when Khomeini had gained power his oscillations can be seen as a way (largely successful) of managing his broad clerical coalition. For instance, by creating the Islamic Republican Party (through his clerical allies) he provided a forum for dissenting opinions within the clerical establishment, and the development of competing wings also allowed him to practise the tactic of "divide and rule". This necessitated alternating support for the different wings to prevent any single branch of the party becoming dominant (Milani, 1997:86–7).[15]

In government, conservative elements in the clerical establishment, centred on the secretive *Hojatieh* society, used their control of the Council of Guardians to block attempts by the Khomeini-inspired Islamic radicals in the Majlis to institute populist economic measures.[16] "A bill proposing only a moderate act of rural redistribution involving uncultivated land was held up as un-Islamic by the Council of Guardians from 1980 onwards" (Owen, 2000:95).

Despite widespread participation of secular leftist and the liberal Islamic republican groups in the early days of the revolution at the time of the Bani-Sadr provisional government, these groups were ruthlessly suppressed by the dominant conservative groups once the latter had consolidated their control over the Islamic Republic's institutions in the early 1980s (Ehteshami, 1995:10–12). The Tudeh, composed of Moscow-leaning communists, was the last secular leftist party to be banned, in 1983.

The structure of the political system (with the existence of parallel institutions like the revolutionary courts) and the tendency towards factionalism (a continuing feature of the Iranian establishment) required a strong leader to centralise decisionmaking. This conveniently served Khomeini's disposition to become ruler of the Iranian state. His vascillation between different factions did, however, feed back into the system creating instability and political log-jam.[17] Owen summarises the situation succinctly as follows:

> Given the separation of powers and existence of so many parallel institutions like the revolutionary courts, cohesion was only possible when implemented by Khomeini himself as the Faqih and leader of

> the revolution . . . [I]n practice, the ayatollah was slow to come down on one side or another in the various factional disputes that ensued (Owen, 2000:95).

The log-jam was marked in practice by events such as the Council of Guardians' opposition to laws passed by the Majlis on the grounds that they were un-Islamic. Ayatollah Khomeini continued to vacillate from one position to another until his death in 1989 (Behdad, 2000:104). Furthermore, the endemic factionalism of the Islamic Republic's leadership and the ambiguity of Khomeini's own position meant that the First Republic leadership (1979–89) failed to address the tension between the material, the political and the spiritual. The regime's legitimacy, derived from popular support, coupled with the exigencies of the Iran–Iraq war and the charisma of Khomeini ensured the continuance of the regime despite its failures. But the scene was set for further failures in the Second Republic.

External Relations

The complex mix of domestic and foreign policy is evident during Khomeini's rule. Iran's position and reputation in the world would be determined by two main events: the 1979–81 hostage crisis and the 1980–88 Iran–Iraq war. These events would cause massive economic and social dislocations within Iran, in addition to wider international hostility (e.g. soured relations with the United States for a generation) and economic difficulties resulting from declining oil prices and significant capital flight. This chaotic economic and political context made the adoption of a coherent policy agenda difficult, even for a skilled political operator like Khomeini (Rahnema and Nomani, 1990:269).

The fact that Iran was now a revolutionary country domestically spread the belief amongst Western countries that its revolutionary impetus was aligned with its foreign policy. The revolutionary rhetoric of Khomeini shaped relations with the West and this has reverberated into the present. In other words, the perceptions about Iran as an international actor are traced to this particular period. They continue to play an important role in shaping the possible options open to Iran internationally, and also domestically, in terms of policies.[18] The hostage crisis (Tehran, November 1979) reflected the different domestic trends with respect to foreign relations issues, and how some factions could act irrespective of the other views existing in the same political party. It put into practice Khomeini's repeated calls for attacks on US and Israeli interests. In Katouzian's words, the hostage taking

> had been either intended or was quickly recognised as a potent weapon for promoting the interests of both the ideological (*maktabi*) Islamists and the Marxist-Leninist left . . . at the expense of more pluralistic trends within the revolution (Katouzian, 1989:57).

The rift grew with the Iran–Iraq war.[19] Both the war and hostage crisis were linked to the political competition within the clerical establishment over the appropriate foreign policy stance. For many Maktabis, the export of the revolution and confrontation with the West were essential parts of the revolutionary programme. "Within the ruling elite, the issue of the war was inextricably linked with the question of the export of the revolution" (Ehteshami, 1995:23). The Iraqi invasion of Iran, although unwelcome, provided an opportunity to export the revolution to Iraq, whose majority Shi'a population, according to Tehran's thinking, might have been open to Iran's revolutionary agenda. This helped to explain Iran's unwillingness to enter into peace negotiations, despite the war's mounting economic and human cost. Khomeini was reluctant to accept the UN Security Council Resolution (SCR 598) demanding an end to the war; he eventually did so only because of mounting pressure from Rafsanjani and other pragmatists.[20]

Similar thinking motivated the arming of Shi'a guerrillas in Lebanon and Afghanistan, the steady criticism of the Saudis' pro-Western policies and Khomeini's final contribution to Iran's foreign policy – the 1989 fatwa calling for the murder of the British author Salman Rushdie. The fatwa provided "a means of meeting [Khomeini's] two main policy goals – mobilisation at home, confrontation internationally" (Halliday, 2003:71). Khomeini's rhetoric in the early days of the revolution had been explicitly internationalist.[21] This populist and nationalist message, casting Khomeini as a "Third World revolutionary" (Roy, 1994:179), essentially died with Khomeini in 1989.

The trade and economic sanctions and the freezing of Iranian assets imposed by the US authorities in response to the hostage taking, combined with the additional economic dislocation caused by the Iraqi invasion towards the end of 1980, eventually led to the release of the US hostages. The hostage crisis and the fatwa against Rushdie contributed, *inter alia*, to a negative picture of Iran internationally and to its diplomatic isolation (Krasner, 1978; 1979). As is discussed in the next chapter, the foreign policy area is one where the victory of "realists" or "pragmatists" (personified by Rafsanjani) since Khomeini's death is clearest (Ethteshami, 1995:29). It was embodied in the first five-year development plan, covering 1990–95, which centred on attracting investment through liberal economic reform. The plan's "emphasis was on economic co-operation with the West and others, which in turn depended on correct, if not warm, diplomatic ties" (ibid.:42).

Economic Interventionism

The stress of having lived through a destabilising and debilitating period during the eight-year war with Iraq was weighing hard on the psyche of the population at large. The rationing of daily requirements was causing unrest amongst the people. Economic policymaking during the 1979–89 period was widely criticised. The stagflation (high inflation and low or negative growth) of the period was a direct result of the authorities' policies. Pesaran puts the economic ills in a nutshell:

> These adverse economic conditions (an acute form of 'stagflation') were due largely to the revolutionary upheavals and their aftermath . . . But they were further exacerbated by the regime's foreign policy adventurism with its adverse consequences for Iran's access to international capital markets, extensive nationalisation of the entrepreneurial and the banking system, continued uncertainties over property rights and the role of the private sector in the economy, centralised and inward-looking government policies aimed at maintaining a highly over-valued official exchange rate through import compression, foreign exchange restrictions and generally interventionist economic policies with far-reaching implications for resource allocation, particularly in the financial and industrial sectors. The result had been an economy in a state of acute disequilibrium with highly distorted prices signals. It was clear that the economic policies of the previous decade could not be continued, and a new approach to the management of the economy was needed (Pesaran, 2000:66).

The economy certainly performed badly over the period.[22] Estimates that real per capita output declined by 4% per year between 1978 and 1988 (Pesaran, 2000:64). Average inflation over the same period was around 18%, resulting in a similar growth in the exchange rate premium (the gap between the currency's official value and its value on the free market). Underpinning the economic deterioration was a collapse in investment.[23]

There were continuities in the economic policies of the Shah's regime, especially with respect to import substitution industrialisation. Thus, while in other developing countries, their planned economies adopted export promotion strategies in favour of import substitution, "in Iran, by contrast, the old regime's import substitution industrialisation policies were retained and intensified into the post-revolutionary period" (Hakimian and Karshenas, 2000:37). One element of novelty refers to

> the greater involvement of the public sector in industrial production, and the dominance of semi-public organisations in the form of various charitable foundations with strong political ties to the regime (such as the Foundation of the Oppressed and Disabled) . . . [T]his ensured the continued subsidization of the old manufacturing sector and the allocation of foreign exchange to old enterprises at highly undervalued official rates at a time of severe foreign exchange shortages (Hakimian and Karshenas, 2000:37).

Turning to other problems resulting from the war with Iraq, Kanovsky writes:

> The 1980–88 Iran–Iraq war, the decline in oil prices over much of the 1980s and the massive capital flight and brain drain following the revolution also played a significant role. The exodus of physical and human capital has been felt particularly strongly. Brain drain was also a major problem as around two million Iranians, mostly highly skilled, left the country during the revolution and its aftermath. This loss is still instantly apparent in present-day Iran (Kanovsky, 1997a:6).

However, policy errors contributed as much to Iran's poor economic performance over the period. Not only were they bad for the state of the economy, but they encouraged already existing practices of corruption and speculation.[24]

Moreover, the economy was hamstrung by the ideologically driven political log-jam. Nomani and Rahnema explain: "This was mainly due to factional disputes over the orientation of the 'Islamic economy' among the rulers of the Islamic Republic" (1994:182).

A five-year development plan was put before parliament in 1982, but implacable conservative opposition within the clerical establishment saw the plan quietly dropped in 1983. In its place, economic policy was driven by short-term and local contingencies and, paradoxically, by institutional inertia which saw many of the Shah's much-criticised development policies maintained, such as import substitution. With large-scale state ownership stymied by conservative opposition,

> . . . the expanded government role . . . was thus not solely or even primarily through a shift in balance from private to public ownership. It was manifested in direct interventions in the operation of markets – foreign exchange controls, maintenance of a system of multiple exchange rates, control on interest rates and bank credits – as well as direct price controls in a large number of product markets (Hakimian and Karshenas, 2000:34–5).[25]

Economics and Islam

As previously mentioned, the issue of private property became an area of particular contention for Iran's clerical leadership. Opinion differed significantly, reflecting the broad coalition of revolutionary forces assembled by Khomeini prior to 1979 and the competing Shi'a subsystems which provided for widely differing interpretations of Islamic law on the issue. There is no conclusive guidance in Islam; "everyone pulls in a different direction according to taste" (Rahnema and Nomani, 1990:269). This doctrinal vagueness has been widely identified as a core weakness in attempts to build up an Islamic government.

> While Islam promises social justice to the poor and destitute, it also makes it clear that the quest for worldly possessions, the accumulation of property and profit, is not only accepted but highly respected . . . The Islamic revivalist movements promise the establishment of a social order that will unite conflicting aims: the welfare of the poor and the security of property (Rahnema and Behdad, 1995:12).

In fact, there is a much deeper contradiction between popular sovereignty and Islamic government in general. For example,

> . . . many Islamic precepts, such as those dealing with the position of women in family and public life, the limitation of the rights of non-Muslims, and certain aspects of social life – for example, Islamic forms of punishments (*hodud-shar'ia*) – are explicitly addressed in the Qu'ran or by Mohammad's Tradition, and are therefore irrefutable. On these and similar matters, there is little room for variation in interpretation (Rahnema and Behdad, 1995:11).

Thus, strict adherence to Islamic precepts might frustrate policies that might be in the interest of the population. Laws that are regarded as necessary by parliament – such as land redistribution, nationalisation of foreign trade, labour legislation – can be rejected on grounds of incompatibility with the Islamic precepts by the Council of Guardians (Rahnema and Behdad, 1995:12–13).

To reach a political compromise on the issue was impossible. As detailed above, factionalism, irreconcilable aims of the conservative and radical wings of the clerical establishment, and the vagueness of Khomeini's guidance led to a situation in which all promised objectives could not be fulfilled. This inability to develop a coherent economic programme may be described as "the greatest failing of Islamist movements in general" (Halliday, 2003:67).

Nationalisation

An issue intrinsically related to the discussion of property rights and to the theme of this study is the balance between national and private ownership in the economy. While the concept of private ownership was not seriously questioned, much of the Iranian economy was nationalised after the revolution (personal communication).[26] This issue is, to some extent, an extension of the confusion over the nature of an Islamic economy, as illustrated above. Islamic principles recognise private property, but there are certain limits on it. In particular, there is a need for some redistribution of wealth. The political incantation of this debate was represented by the Hojatieh faction, which leaned towards lower taxes and a more free-market approach, while the Maktabis leaned towards centralised economic planning. This debate should be seen in the context of political factionalism. However, one should remember that there are many groups within each of these two groups having different shades of opinion; for instance, followers of the Hojatieh do not always concur with many of the Hojatieh doctrines, and, similarly, not all Maktabis are committed supporters of centralised economic planning.

Before looking at nationalisation in general, it should also be noted that there was more at stake than just a shift from private to public ownership. As described above, especially in the quotations of eminent analysts, direct interventions in the operation of the market had "serious consequences at all levels of economic activity from investment to production, trade, distribution, and consumption" (Hakimian and Karshenas, 2000:34). The constitution of 1979 divided the economy into three sectors: state, co-operative and private. To be included in the state sector were strategic industries, such as mining, the railways and foreign trade. In particular, as was the case under the Shah, the state had full control over oil revenues. The private sector was secondary to the state sector and included industry that complemented the economic activity of the state as well as agriculture-related business.

Formally, the areas of nationalisation comprised properties belonging to the Shah's court, of owners who had fled the country, and firms that owed more than half of their assets to banks (ibid.). The nationalised industries consisted of those that had been controlled by the court elite, including Iran's 51 major industrialists. This led to the state's control of over 80–85% of the country's industry (Ehteshami, 1995). The court's holding company was transferred to a new foundation, the Foundation of the Deprived. Later, more Islamic foundations were formed to act as additional holding companies. Banking, insurance and mining were also fully nationalised.

Where expropriation of private assets did occur, it generally happened on an *ad hoc* basis in response to local imperatives and conditions. Many of the industrial concerns nationalised by the Islamic Republic were heavily in debt to state banks, so that "their nationalisation was largely a method of cancelling their debts to the state" (Katouzian, 1989:56). Others were expropriated in spontaneous takeovers by workers' groups (from the autumn of 1978 onwards), typically after the owners had fled Iran (Behdad, 2000:101–2). The institutions of Khomeini's mini-state were often involved. The Revolutionary Courts confiscated assets of "anti-revolutionaries"; the assets were transferred to the newly created Islamic foundations, of which the Foundation for the Oppressed (Mostazefan Foundation), the Martyr's Foundation and the Fifteenth Khordad Foundation became the most significant. This early phase of expropriation was endorsed and sometimes promoted by Khomeini's regime "as a means of mass mobilization" (ibid.:102). However, once power was consolidated, such expropriations fell away and the *ad hoc* nationalisation programme was largely halted.

It is possible to discern three broad trends which defined the level of nationalisation in the post-revolution economy of Iran. Firstly, the debate over a populist redistribution of wealth led to considerable nationalisation. Secondly, the mounting failures of Iran's economy led to further nationalisation to bail out other industries. Finally, nationalisation was, to a certain extent, a transfer of previously nationalised property from one government to the next. The court's property simply became the new government's property and that of its Islamic foundation appendages.

The populist leanings of the government necessitated some redistribution of wealth, i.e. nationalisation. In order to mobilise the people, Khomeini needed to make promises about an equitable economic system and so it was necessary to carry out at least some of these promises when the revolution had finally succeeded. (In this connection, the Latin American populist revolutions come to mind.)

Ehteshami makes the case that the nationalisation of industry, rather than being an attempt to revolutionise the economy, was in fact an attempt to "bring back productivity as well as order and management to a large and abandoned section of the economy" (Ehteshami, 1995:86). Thus the reasoning behind nationalisation is partially attributable to the personal intervention of Khomeini who was anxious to avert an economic collapse. In other words, the state was bailing out the failing industrial sector by means of formally nationalising firms with large debts. Alternatively, it is possible to see the nationalisation of economic sectors in terms of serving a similar goal as it did under the Shah (personal communication).[27] As demonstrated in Chapter 2, the Shah used his control of the economy, especially of oil revenues, to increase his power and dominance over society. In a similar manner, the post-revolutionary government also used its control of the economy to further its power and dominance over society. By controlling large industries – points at which it is possible to accumulate power – the post-revolutionary government was able to limit possible counter-revolutionary activity, especially in the early years.

Nationalisation extended to the financial sector. Like many developing economies, the financial sector was subject to repression – government controls and policies which suppressed market forces and directed credit to favoured groups, a clear case of clientelism. The provisional government had nationalised some banks in 1979 (Behdad, 2000:101–2), while the authorities introduced Islamic banking through the Law of Usury-Free Banking, approved by parliament in August 1988 (Nomani and Rahnema, 1994:178). The financial sector's performance deteriorated over the period, a result of uncertainty about government regulations – with "rates of profit" (interest rates in an Islamic guise) well below market clearing rates (and well below inflation) – and, not least, from poor management and red tape. Monetary policy was subject to "fiscal dominance" as the large government deficit was "financed by borrowing from the Central Bank and the banking system" (ibid.:181–2). The inevitable result was spiralling inflation and a plummeting currency. The authorities' attempts to offset these effects via direct price controls and multiple exchange rates only served to further economic instability.

The Bonyads

The formation of the large Islamic foundations (Bonyads) was one of the novelties that sprang from the regime's economic policy, but unfortunately they too became a vehicle for dispensing favours to the regime's political loyalists. These semi-public bodies under clerical control have come to dominate the Iranian economy and many also have significant asset holdings outside Iran.[28] The Bonyads were the chief recipients of the distorted prices created by the regime, gaining access to scarce foreign exchange at subsidised prices. The foundations, essentially networks of conglomerates, were controlled by an emerging "entrepreneurial" class "who were close associates of the regime and benefited from the many privileges of this association" (Behdad, 2000:113).[29] Kanovsky is highly critical of the Bonyads' role in the Iranian economy, arguing that although their stated purpose is to help the poor, they helped themselves. He writes:

> Profits from [their] enterprises to provide inexpensive housing, healthcare, and other social services to the poor, [in reality much of the profits is] siphoned off by those in control and relatively little reaches the needy. Though the Bonyads' 1992–93 budget was equivalent to nearly half of the government's budget that year, the latter exercises very little control over these foundations, which answer only to the supreme leader [*faqih*] (Kanovsky, 1997a:7).

In personal interviews with the Chairman and Managing Director of the Audit Organization,[30] it was discovered that the Bonyads own:

- 140 industrial enterprises, manufacturing products such as food and beverages, electrical appliances, construction materials, textiles and clothing, leather goods, plastics, metals and paper;
- 64 mines;
- 2,786 properties, covering agricultural, commercial or construction sites;
- 230 commercial and trading companies;
- 90 theatres and three of Tehran's leading newspapers.

They also own a hospital, retail outlets, airline and shipping companies, an insurance company and a credit institution. In addition, they are involved in the construction of commercial buildings, roads and railways. They employ some 150,000 people and produce 28% of Iran's textiles and clothing, 22% of its concrete, 45% of its beverages, 28% of its rubber products and 25% of its sugar. Its annual budget is reported to be around $10 billion.

The concentration and monopolisation of financial power in the hands of the Bonyads provide them with a disproportionate economic weight in relation to any single private firm wishing to obtain a share of any of its markets. Their decisions concerning production and imports have major consequences for the availability of raw materials, market prices of products and the profit margins of competitors. The country's economy would benefit greatly if these establishments were to be broken up, regularised and eventually privatised so that a competitive and efficient business environment could emerge in Iran.

Oil Policy

The oil sector continued to dominate the Iranian economy during the 1979–89 period, although oil revenues were severely reduced by the collapse of the oil price in the mid-1980s – oil revenues fell from $21 billion in 1983 to $6 billion in 1986. The US trade embargo and the Iran–Iraq war (particularly the 'tanker war') further damaged an already weak economy (Kanovsky, 1997a:10). With respect to the oil sector policy, the authorities brought all the oil and gas industry's activities under the auspices of NIOC, cancelling the agreements with the international consortium and reducing the production target (Rahnema and Nomani, 1990:240). The reduction in production reflected the pre-revolutionary rhetoric against the Shah's oil policies, which Khomeini accused of favouring Western imperatives (ample, cheap oil). As in other areas of the economy, clerical opposition to technocrats (perceived as pro-Western) led to a significant brain drain which in turn robbed the sector of management expertise. As early as the late 1980s, the regime was forced to reverse its initial decision to cut oil production, as the spiralling costs of the Iran–Iraq war coupled with the economic crisis caused by the dramatic fall in oil prices took their toll (Richards and Waterbury, 1996:57–9). Since 1989, this renewed emphasis on oil and gas exports (yet another return to the Shah's development strategy) has become more pronounced.

Interestingly, while the government's control over oil revenue allowed it to gain dominance over Iranian society, it forced the government to develop and maintain relations with the international capitalist system. This was necessary in order to secure the sale of oil and gas (personal communication).[31] In particular, it was necessary to work with the West to some degree, even when suffering from a Western embargo. While there was an attempt to move towards economic self-sufficiency, oil sales could never truly be replaced. Thus, whether or not the post-revolutionary government desired to isolate itself from the international system, it could not possibly do so if it wanted to retain governance of Iran. The overall deterioration of the economy, especially during the 1980s, prompted a change in economic policy. In particular, the government now sought more private sector involvement and formally announced this in 1986 (Ehteshami, 1995). The 1986/1987 oil crisis also forced the government to increase taxation in order to continue paying for the promised subsidies.

Economic liberalisation continued after Mir Hussein Moussavi's re-election as prime minister in 1988. The government lifted some restrictions on the importation of certain goods. The Tehran stock exchange was reopened in the autumn of 1988. Earlier that year, the government announced a comprehensive privatisation plan in which "non-essential and non-strategic" industries would be privatised (ibid.:97). After much debate, by 1989, the government had reversed its long-standing policy against foreign borrowing and decided to borrow $2 billion over the subsequent five years (ibid.).

The deep-rooted and self-inflicted problems of Iran's oil policy stem to a great extent from a number of factors: prolonged US sanctions; the Islamic Republic's ideological bent and populist policies; and the country's increased isolation from the Western diplomatic community in recent years. These factors have compounded the problems of supply, which are due to the nature of the buy-back contracts, driving many potential investors away. Furthermore, with rising domestic demand for goods and a rising population, not to mention the various subsidies that the government was committed to provide, one can say that Iran's economy was in dire straits.

Clearly, the Islamic Republic under Khomeini plunged the country into chaos. Popular sovereignty and theocracy did not blend as had been hoped; nationalisation of industry and financial institutions led to inefficiencies, price distortions and multiple exchange rates; expropriated property was awarded to the Bonyads, sprawling conglomerates, posing as charitable foundations to help the poor, came to dominate the economy – with the blessing of the clerical regime – and nearly brought the economy to its knees. They are seen in Iran as massive corruption machines. Factionalism, cronyism and clientelism were rampant; relations with the West crumbled.

With the death of Khomeini in 1989, the leadership passed to more pragmatic and reformist figures such as Ayatollah Ali Khamenei, the spiritual leader, and Ali Hashemi Rafsanjani, who was elected Iran's first Executive President. During Rafsanjani's presidency, Khomeini's radicalism was replaced by a willingness to compromise, particularly over economic policy. A programme of economic reform was launched and a policy of détente was adopted in foreign affairs.

NOTES

1 Personal communication. Discussions with Dr Bijan Bidabadi, Faculty Member & Director of Foreign Exchange Research Department of the Central Bank of Iran, Tehran, December 2005.

2 Abrahamian (1991) notes that the constitution contains significant populist elements, yet also undertakes to safeguard private property, while Owen (2000) draws attention to the disjoint between its secular and Islamic features.

3 This emphasis on clerical interpretation is also present in the Shi'a subsystems associated with Shariati, Motahhari and Navab-Safavi (as discussed in Shahriari-Rad, 2002).

4 Personal communication. Discussion with Dr Manouchelir Lighrani, M.D., former landowner, Tehran, December 2004.

5 He was assassinated in May 1979 during the bloody early months of the Islamic Republic.

6 Unlike Milani, Owen argues that the elimination of opposition forces and alternative centres of power was undertaken by the Islamic Republican Party leadership rather than by Khomeini directly, as "the role of the Ayatollah Khomeini . . . was not at all clear cut" (Owen, 2000:180).

7 Milani argues that factionalism was a new development under the Islamic Republic. However, the Shah's attempts at creating civil society institutions in the second half of the 1970s, discussed in the previous section, constitute the first flowerings of factionalism. Milani is undoubtedly right, though, to contend that "it is because of factionalism that some primitive kind of pluralism among the governing elites is developing in Islamic Iran . . . The supporters of the Islamic Republic enjoy more freedom to debate important issues than the Shah's supporters had under the monarchy" (Milani, 1997:86).

8 The conservatives also tended to be less inclined towards Khomeini's thesis of '*velayat-e faqih*', or jurist's guardianship, his conception of Islamic government that placed the clergy at the centre of power and allowed them to interpret Islamic law indirectly rather than impose *sharia* directly (Wells, 1999). As Roy argues, "Khomeini always imposed revolutionary logic, represented in the guide's will, if need be over the *sharia*" (Roy, 1994:176). Conservatives, by contrast, tended to favour direct implementation of *sharia* rather than the essentially secular legal framework, combined with *velayat-e faqih*, created by the framers of the 1979 constitution. This conception of Islamic governance was actually the mainstream view amongst the upper echelons of the clergy, even in the early 1980s (ibid.:173). The succession of Khamenei, a relatively junior cleric and hardly the senior jurist demanded by the logic of Khomeini's philosophy, to the *faqih*-ship significantly undermined the system's authority.

9 Personal communication. Discussions with Dr Bijan Bidabadi, Faculty Member & Director of Foreign Exchange Research Department of Central Bank of Iran, Tehran, December 2005.

10 However, some analysts put it more strongly. Rahnema and Nomani note: "Infighting has been endemic to the Islamic Republic in the process of its formation and establishment and it has its roots in the different politico-economic interests of the clergy and the realities of a crisis-ridden Iranian society. *Ad hoc*

decisionmaking and reaction to short-run economic events, political conflicts, rhetoric and phrasemongering, confusion and uncertainty and the existence of semi-anarchic decisionmaking centres prevented the implementation of a stable and consistent programme of economic reform between 1979 and 1989. Measures that have been implemented have not been sufficient or efficient enough to check the disastrous economic downturn which has characterised this period. In 1984, a clerical member of the second Islamic parliament summarised the fate of the economic institutions and policies of the Islamic Republic: 'Islam has a definite economic system [but] since it has not been fully compiled, everyone pulls it in a different direction according to taste.' This situation has resulted in a severe and protracted economic crisis and a lower standard of living for the Iranian people" (Rahnema and Nomani, 1990:269).

11 Milani also cites Ayatollah Mohammad Hosseini Beheshti, the writer of the *velayat-e faqih* provision of the constitution, who argued that once the Iranian people had freely embraced Islam in the April 1979 referendum on the creation of the Islamic Republic, then the limits of popular sovereignty must be dictated by the boundaries of Islam, interpreted by the clergy (Milani, 1997:83).

12 Personal communication. Discussions with Dr Bijan Bidabadi, Faculty Member & Director of Foreign Exchange Research Department of Central Bank of Iran, Tehran, December 2005.

13 Personal communication. Discussion with Dr Manouchelir Lighrani, M.D., former landowner, Tehran, December 2004.

14 Personal communication. Discussions with Dr Bijan Bidabadi, Faculty Member & Director of Foreign Exchange Research Department of Central Bank of Iran, Tehran, December 2005.

15 Khomeini eventually tired of the party's squabbling and dissolved it in 1987.

16 Ehteshami details the structure and ideological orientation of the various factions making up Iran's clerical establishment in the post-revolution period (Ehteshami, 1995). The *Hojatieh*, initially formed in the 1950s as an anti-Bahai organisation, was fiercely anti-communist, socially traditional and supportive of free enterprise.

17 Personal communication. Discussions with Dr Bijan Bidabadi, Faculty Member & Director of Foreign Exchange Research Department of Central Bank of Iran, Tehran, December 2005. See also Chapter 5 for a discussion of the Islamic Republic's political institutions.

18 For example, see the policies on nuclear energy in Chapter 5.

19 The Iran–Iraq war was precipitated by the Iraqi attack on Iran in September 1980. Iraq was widely perceived to have obtained at least tacit approval of its invasion from the West, the United States more specifically.

20 Khomeini stated that "the acceptance of the resolution . . . was a bitter and tragic issue for everyone and particularly for me . . . death and martyrdom would have been more bearable to me" (Ehteshami, 1995:28).

21 "The problems of the East come from the West – especially from American imperialism . . . the oppressed nations of the world should unite against their imperialist oppressors" (Abrahamian, 1991:114).

22 Another analyst points to yet more reasons: "The clerical regime used the war to explain all or most of the country's economic problems. In reality, many stemmed from the distorting and destructive policies adopted after the revolution. The complex system of multiple exchange rates, price controls and rationing, import

licenses, and various direct and indirect subsidies distorted resource allocation and adversely affected production and productivity . . . Moreover, these economic policies greatly increased opportunities and inducements for bribery and corruption. Rationed goods sold at controlled prices were often 'diverted' by merchants and corrupt officials to the free market, where they were readily available at far higher prices. Similarly, when combined with a foreign exchange allocation at an artificially cheap exchange rate, an import license amounted to 'a license to print money'. These kinds of activities were far more lucrative than investment in industry or other productive sectors" (Kanovsky, 1997b).

23 Private investment by 1988 was less than a quarter of the level in 1977 (Kanovsky, 1997a, p. 53), and much of the investment that did occur was driven by cronyism and distorted price signals, and was probably therefore of doubtful economic value.

24 Kanovsky's book was published by the Washington Institute for Near East Policy, whose board members included leading neoconservatives such as Richard Perle and Paul Wolfowitz, noted for their hostility towards the Islamic Republic of Iran. However, notwithstanding this potential source of bias, Kanovsky's criticisms of Iran's economic policies over the period are in keeping with those of others.

25 Hakimian and Karshenas (2000) provide evidence of the price distortions introduced by these policies: the black market exchange rate premium reached more than 2000% by the end of the 1980s, while the prices of energy, bread and other key products had fallen to well below 10% of prices prevailing in the international market.

26 Personal communication. Discussion with Dr Manouchelir Lighrani, M.D., former landowner, Tehran, December 2004.

27 Personal communication. Discussions with Dr Bijan Bidabadi, Faculty Member & Director of Foreign Exchange Research Department of Central Bank of Iran, Tehran, December 2005.

28 Behdad (2000:113) claims that the Mostazefan Foundation is "the largest economic entity in the Middle East".

29 The Bonyads' economic structure (sprawling conglomerates), their client relationship with the ruling class and their dominance of the domestic economy make them close relatives of the Korean *chaebol*. Their potential for generating macroeconomic crises – similar to the *chaebol*'s role in the 1997–8 crisis in Korea – should not be underestimated.

30 Interview with Mr Houshang Naderian, Chairman & Managing Director, Audit Organization, Tehran, December 2005.

31 Personal communication. Discussions with Seyed Mohammad Hossein Adeli, former Iranian Ambassador to the UK and present editor of *Economic Trends*, Tehran, May 2006.

4

POLITICS AND ECONOMICS AFTER KHOMEINI

Introduction

This chapter builds on the previous chapters dealing with the history of the political economy in Iran, the purpose of which is to come closer to answering the question of whether there is the political will to privatise oil and gas in the country. It addresses several major sub-questions. The first is: what are the determinants of the Iranian policy of oil and gas privatisation in Iran? As with the previous chapters, this question is answered by exploring the internal politics, such as the role of leaders, the political make-up, external relations and economic policies. The second question is: which are the most influential constituents in the process of formulating the privatisation policy? Again, the chapter addresses the issue by examining the major institutional constituents that *determine* the policy. These include the executive and legislative branches of the republic, society as a whole, the media and religious elements. These constituents will be examined within the framework used in previous chapters. But before tackling these questions, it may be useful to begin with a brief history from 1989 to the present period so to provide a frame of reference for the rest of the chapter.

The historical period that this chapter focuses on has seen significant internal political changes which made a noteworthy impact on the prospects of privatisation in Iran. During that period, four presidents came to power, three of whom left their mark in a significant way. Important changes in the Iranian political structure took place in the late 1980s and early 1990s.[1]

As mentioned before, in August 1989, Ali Akbar Hashemi Rafsanjani, the longtime speaker of the National Assembly, was elected president with an overwhelming majority of the vote and Ayatollah Ali Khamenei assumed Khomeini's role as spiritual leader or *faqih*. Nateghnour won the election in 1993, but with a modest vote compared to Rafsanjani's and in 1997 he lost to the moderate reformer, Mohammad Khatami who was subsequently re-elected, in 2001. He tried to restore normality in Iran and in its relationships with foreign countries. Elected with 69% of the vote in the May 1997 elections, and again, with some 60% of the vote, in the 2001 elections, he certainly had a clear mandate from Iran's youthful population.[2] However, in 2005, the conservative mayor of Tehran, Mahmoud Ahmadinejad was elected President, swinging the pendulum of reform back to conservatism.

Internal Politics

The domination of foreign powers in Iran's modern history had a profound effect on the mindset of the political elite. The fear of being dominated again was etched into the memory of the people to the point where they erupted in 1979. The clerical establishment never forgot the years of foreign manipulation and interference in the internal politics of modern Iran. Perhaps it would not be too difficult to understand the animosity and apprehension the decisionmakers felt towards the foreign players. The "Great Game" that Russia and Britain played for control of Central Asia in the nineteenth century left scars on generations of Iranian leaders and instilled in them a determination to insulate Iran from foreign influence. This long-standing suspicion developed into a deep ambivalence towards foreign powers within the modern clerical establishment. To this day, the internal politics of post-revolutionary Iran is shaped by this sentiment.

The leadership structure

Khomeini was a strong and charismatic figure in control of the state apparatus. His personal beliefs impacted on the direction of policies for a decade. The following analysis aims to identify the role of the subsequent leaders of Iran in shaping policies.

One important issue that needs to be taken into account after the death of Khomeini is the more marked separation of the institution of the spiritual leader from that of the President. The latter's role has been (at least legally) strengthened with enlarged powers under the amended post-Khomeini constitution. For example, the post of Prime Minister was abolished and some executive powers were transferred to the presidency, creating an uncomfortable partnership between the elected President and the supreme leader. Since 1989, the President has appointed the cabinet, although all ministers must be approved by the Majlis before taking office. The supreme leader still enjoys primary control over many organs of state and retains the right to appoint key officials such as heads of the judiciary, the media, the armed forces and various revolutionary bodies (Anon, 2005b). Thus, the lack of an overlap between the political leader and the spiritual one is an important aspect of Iranian politics since 1989. However, although the *faqih* still retains considerable influence, having the power to intervene in any issue, including specific pieces of legislation, the approval of the final version is more about bargaining (as in the case of political bureaucracies in secular, democratic states) rather than outright imposition of the *faqih*'s views.

The religious figure who has been most influential in the Iranian political environment is Ayatollah Ali Khamenei. He was appointed supreme leader for life in June 1989; he was one of the founders of the Islamic Republic Party in 1979 and served under Ayatollah Khomeini fulfilling different functions.[3] He also served two consecutive terms as President in the 1980s.[4] Khamenei's influence is seldom exercised explicitly, often behind the scenes. He has aimed to preserve stability and to contain the power struggle between reformist and conservative factions, but as the schism within the political elite has widened he has increasingly backed the conservatives (Anon, 2005b). In other words, Khamenei, although not formally allied to this group, is viewed as being broadly supportive of the conservative position on most issues. Over the years, he has come into conflict with the former Presidents Ali Hashemi Rafsanjani and Mohammad Khatami, as well as with other reformists (Wells, 1999:30; Owen, 2000:112; BBC, 2005b).

Rafsanjani, elected President in 1989, is a figure whose affiliations are harder to place. Owen places him in the reformers' camp; a third faction made up of "moderates" (social liberals with "third way" economic leanings), former radicals and pragmatic careerists (Owen, 2000:112–3).[5] Abrahamian (1991; 2001:119) characterises Rafsanjani's politics as closer to the conservative position and essentially similar to Khamenei's, particularly in the field of economic policy. Ehteshami argues that, at least initially, Western analysts tended to place Rafsanjani in the Maktabi camp, but that latterly he (and Khamenei) have been given a subset of their own – the "technocratic coalition" (Ehteshami, 1995:18). However, Ehteshami holds that this group did not form a "homogeneous and cohesive force", but rather attempted to forge a pragmatic approach to policy formulation that "led to a sense of insecurity among the governing ranks of the republic, which in turn bred opportunism within the elite" (ibid.). Rafsanjani now enjoys a position of power and influence as the head of the Expediency Council, from which, according to Owen, he has ensured "that attacks on [his successor as president] Khatami's leadership could be blunted and contained" (Owen, 2000:113). This view supports the proposition that Rafsanjani's political instincts are essentially reformist.[6]

During his first term of office President Rafsanjani reinforced his reputation as a corrupt and power-hungry politician (personal communication).[7] He crushed personal freedoms and presided over a sharp economic downturn. He ushered in a particularly aggressive phase of Iranian sponsorship of terrorism, including alleged roles in the bombing of a Jewish community centre in Buenos Aires in 1994 that killed more than 80 people, and in the assassinations of several Iranian exiles, including former Prime Minister Shahpur Bakhtiar in 1991.

Khatami, who was elected President in 1997, belongs to the relatively small, moderate wing of the clerical establishment. He had enjoyed a ten-year ministerial career in the 1980s and early 1990s, but had fallen foul of the then dominant conservatives in 1992. He was a supporter of Ayatollah Montazeri, who was Khomeini's designated successor as *faqih* up until March 1989 (Wells, 1999:36). Montazeri fell out with his former mentor after criticising the regime for the slow pace of political reform and its autocratic leanings, resulting in his replacement by Khamenei, a then relatively junior cleric who was hastily upgraded to Ayatollah status (Ehteshami, 1995:33).

Khatami's election campaign focused on the need for political and social reform, calling for the establishment of an "Islamic democracy" (Owen, 2000:113). He drew support from a disparate range of groups, all outside the conservative camp, many of whom the power-brokers of the Islamic Republic had sought to suppress or ignore in the past: they comprised secularists, Islamic leftists and women (ibid.:112). However, despite a significant electoral mandate in both the 1997 and 2001 elections, the President had to struggle hard to build a working coalition to enable him to implement his liberalising agenda. In large part, this was due to "the fact that the Majles [before 2000] and many of the major organs of government were controlled by his conservative opponents" (Owen, 2000:113).[8] Khatami also pushed for economic reform, and indeed his administration implemented structural reforms in a number of areas (IMF, 2003). This success in the economic sphere contrasted starkly with his failure in the Majlis (before 2004) to advance his social reform agenda. Powerful conservative opposition in the Council of Guardians and the Expediency Council as well as pressure from Khamenei, blocked Khatami's attempts at reform.

Mahmoud Ahmadinejad, the President elected in June 2005, was the former mayor of Tehran. His presidential campaign focused on poverty, social justice and the distribution of wealth in Iran; he also railed against corruption. However, when he was mayor of Tehran, his conservative stance on reforms led to the curtailment of many of the reforms put in place by the moderates who had run the city before him. Despite the fact that he was the first non-clerical president for 24 years, Ahmadinejad's attitude shows that Khomeini's legacy was still strong. His populist tone echoes Khomeini's, and his support in the Majlis reflects this orientation (BBC, 2005c). Although not an extremist, he is rather conservative about a range of domestic issues (BBC, 2005d). In regard to extremism, he said: "On the domestic scene, government policies will be based on moderation and any extremism will not only be avoided, but will be dealt with seriously" (ibid.).

His role in reinforcing the people's hostility towards the Western world (the EU, the US and the Western-dominated UN) has been already tested on the nuclear issue.[9] The suspicion that he was one of the students involved in the hostage-taking in 1979 contributed further to the negative image of the Iranian political establishment. He seems to be in agreement with Khamenei, more so than with the former presidents (BBC, 2005c). However, he has difficulty in imposing his will on the Majlis, as shown by his failure to secure its approval of the appointment of the new minister of the oil industry (Anon, 2005c).

In regard to the *faqih*, one may safely say that he still exercises important power in the decisionmaking process and in legislation. However, the clearer separation (after Khomeini's death) of the institution of the President from that of the religious leader created ample opportunities for reform and expression of alternative reformist views. Presidents like Khatami did succeed in imposing a different type of economic (liberalisation) and social policies, despite many limitations (e.g. the limited role of women). Support from amongst the members of the Majlis, the Council of Guardians and the Expediency Council,[10] is also crucial in the implementation of laws. For clarity, Table 4.1 below summarises the various agencies of formal power in Iran as of 1991.

TABLE 4.1
The Agencies of Power as of 1991

Agency	Founded	Function
Faqih	1979	Spiritual leader
Executive Presidency	1989	Executive leader
Cabinet	1979	Control of ministries
Assembly of Experts	1979	Nominates and chooses the *faqih*
Council of Guardians	1980	Ratifies Majlis legislation and supervises elections
Majlis	1980	Approves government policy
Expediency Council	1987	Arbitrates between Majlis and CoG
Reconstruction Policy Making Council	1988	Formulates reconstruction policies
Head of Judiciary	1989	Oversees courts
Joint Chiefs of Staff	1991	Co-ordinate defence policy

Features of the political system

As discussed earlier, Iran's political complexity comes from a blend of modern Islamic theocracy with elements of the popular sovereignty of democracies. This is reflected in the type of institutions that co-exist in Iranian society: a network of unelected institutions controlled by the extremely powerful conservative supreme leader is countered by a president and parliament elected by the people. The purpose in highlighting this is to demonstrate the multiplicity of the decisionmaking bodies within the government which have varying degrees of political will in reforming the existing institutions in Iran.

Figure 4.1 on the next page illustrates the hierarchical composition of the system and its division into elected and unelected institutions in relation to each other (Buchta, 2000; BBC, 2005b). This separation of power becomes more pronounced after 1989 due to the disconnection of power of the institution of the president from that of the supreme leader.

The fact that there were two centres of power – one elected, the other unelected – accounted for the persistent disjunction between state and society and for the endemic political factionalism. The last two features continued to play a crucially important role in the determination of Iranian politics, as will become clear in the next sections. Such a political landscape was bound to have a significant effect on the political economy of privatisation in Iran. Iranian politics has been moving towards a distinction between conservative and liberal views of the world – a distinction which has arisen not just in Iran but in many nations around the world, particularly in the Middle East. Thus, while the argument here begins by focusing on the continuity with the previous eras that we have explored above, the final conclusions will be based on this conservative–liberal breakdown, one which has affected not only internal politics, but also the arguments regarding foreign and economic policies, especially those related to the potential for privatisation of the nationalised state industries.

FIGURE 4.1
Composition of the Iranian Hierarchy

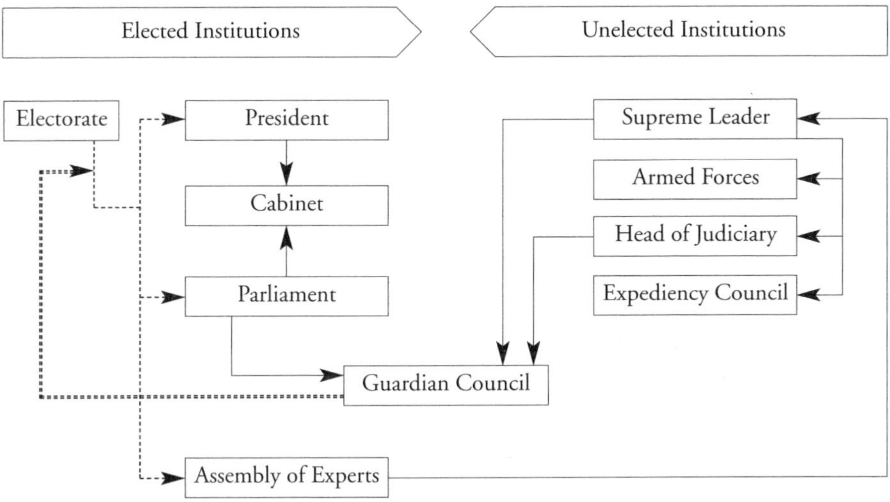

Key: ➤ Appointed of approved ---➤ Directly elected ······➤ Vets candidates
Source: Buchta (2000).

The relationship between state and society

An important element that has become more visible since 1989 is the participation of society in politics. Although often limited to casting votes or sporadically making protestations in the streets, the situation may be considered as an improvement on the previous decades. The spread of information technology and satellite TV has contributed to the creation of a more aware and active civil society. The fluctuations in the turnout at elections reflect Iranian society's attitude towards the affairs of state. In the past, the public's apathy towards politics revealed itself in low attendance at polling stations, but now the electorate has begun to feel that by voting for a specific candidate they would be making a contribution to the shaping of Iran's future.

The lack of political will and the difficulties encountered in the implementation of laws in the Iranian political system, coupled with the continuous conflict between the conservative and reformist camps, can explain why the overwhelming majority that voted for Rafsanjani in 1989 did not come back with the same force in the 1993 elections (personal communication).[11] The gap between the government and society deepened and the disenchantment of the people reached a climax in July 1999. During this month, there were large-scale anti-government protests in the streets of Tehran which lasted more than a week. The police and pro-government vigilantes were sent out to disperse the crowds (personal observations at that time). In a country whose government had only recently came to power through civil unrest, such protests were extremely ominous.

This post-1989 period reflected a change, or at least a recognition on the part of some sections of society of the fact that their views were different from those of the Council of Guardians. In June 2001, Khatami's re-election signalled this fact. One should not assume, of course, a homogenous outlook by society. As with politics, a complex variety of opinions could be found amongst the common people. After Khatami's election, the more conservative elements in the Iranian government invoked the support of the more conservative elements in the country (personal communication).[12] The conflict between the different views was dealt with in the same way as that used by the previous authoritarian regimes of the Shah and Khomeini. Liberal newspapers, representing the opinions of the reformers, were banned. Many of the reformist candidates were disenfranchised. The Council of Guardians enforced discretionary measures to counter people's preferences for different political candidates, as happened in the 2005 elections.

The dire state of the economy, which mostly hurt the lower and working classes, made a populist platform of candidates running for office particularly appealing to them. Ahmadinejad's populist pronouncements attracted those who felt cheated by the trajectory of the economy and had not benefitted from rising oil prices. These social groupings were less interested in concerns such as social liberalisation, which the better off residents of Tehran found important.

Another factor that has come to play an increasingly significant role in reshaping the interaction between society and politics in Iran is demographics. The fact that more than 50% of the Iranian population was now under the age of 18 (personal communication),[13] meant that half Iran's citizens had not been born at the time of the revolution, and therefore had not directly been exposed to the revolutionary discourse, making them more receptive to the ideas of reformists and liberals. Their participation in the 1997 and 2001 elections was thus politically significant.

Factionalism

Factionalism remained a trait of Iranian politics throughout the presidencies of Rafsanjani and Khatami, and at the time of writing the signs so far were that it would continue during the presidency of Ahmadinejad (Owen, 2000; Anon, 2005c). The fascinating feature of Iranian politics is the different combination of socially conservative or reformist agendas and liberal or conservative/socialist economic programmes. Examples of such combinations are briefly presented here. One is the left-leaning faction that favoured redistribution and social justice, variously referred to in the literature as "hard-liners", "Islamic socialists" or "radicals" and known in Iran as the "Imam's Line" or Maktabis (Ehteshami, 1995). They managed to initially rally the greatest electoral support and were well represented in the Majlis, at least until the 1992 elections.[14] The conservative faction, initially centred on the Hojatieh society – later operated through the Society of Combatant Clergymen of Tehran (*Ruhaniyat*) – was aligned with powerful bazaar merchants, advocating a more economically liberal, but socially conservative, agenda.[15] A new coalition was created in opposition to the conservatives as a result of the defection of the moderates and pragmatists from the conservative camp and the increasing pragmatism shown by the former hard-liners in economic policies. This coalition, which can be broadly characterised as liberal,[16] backed Khatami's presidential campaign calling for social reforms.

The election of Ahmadinejad *may* change the factional politics of Iran, but this seems doubtful. Rather, it is more likely that the distinction between "conservative" and "liberal" will come to dominate decisionmaking. However, we may see a rewriting of this distinction along the lines outlined above – i.e. a separation between social and economic issues. The conservatives' success in the 2004 Majlis elections following the blocking of most of the moderate candidates, cemented their victory (Muir, 2004). However, despite the fact that parliament was dominated by conservatives (and continues to be so), many of whom supported Ahmadinejad against Rafsanjani in the second round of the presidential elections in June 2005, the new President has not been able to win the necessary support for decisions concerning the oil industry. It seems that only a minority supports the President (Anon, 2005c).

The persistence of factionalism makes the political environment rather unstable and undermines the decisionmaking process and the implementation of policies. On the positive side, as a form of political pluralism, it allows the expression of an assortment of views. However, as will be demonstrated in the next section, due to the variety of inclinations among the political elite and differing levels of political will, the execution of reforms may be frustrated by the very same individuals who had agreed and approved them. To reach a consensus and bring harmony will be a formidable task.

Table 4.2 and Figure 4.2 show, respectively, the political leadership by name and title, and the country's political structure, as at 2006.

TABLE 4.2
Iran's Political Leadership, 2006

Name	Title
LEADERS:	
Ayatollah Ali Khamenei	Supreme Leader (*rahbar*)
Mahmoud Ahmadinejad	President
Ali Akbar Hashemi Rafsanjani	Chairman of Expediency Council
KEY MINISTERS:	
Parviz Davudi	First Vice-President
Mohammad-Reza Eskandari	Agriculture
Masoud Mir-Kazemi	Commerce
Mohammad-Hossein Saffar-Harandi	Culture & Islamic Guidance
Mostafa Mohammad-Najjar	Defence
Davood Danesh-Jafari	Economy & Finance
Ali-Asghar Fani	Education
Parviz Fattah	Energy

Source: Economist Intelligence Unit (2006).

FIGURE 4.2
Iran's Political Structure, 2006

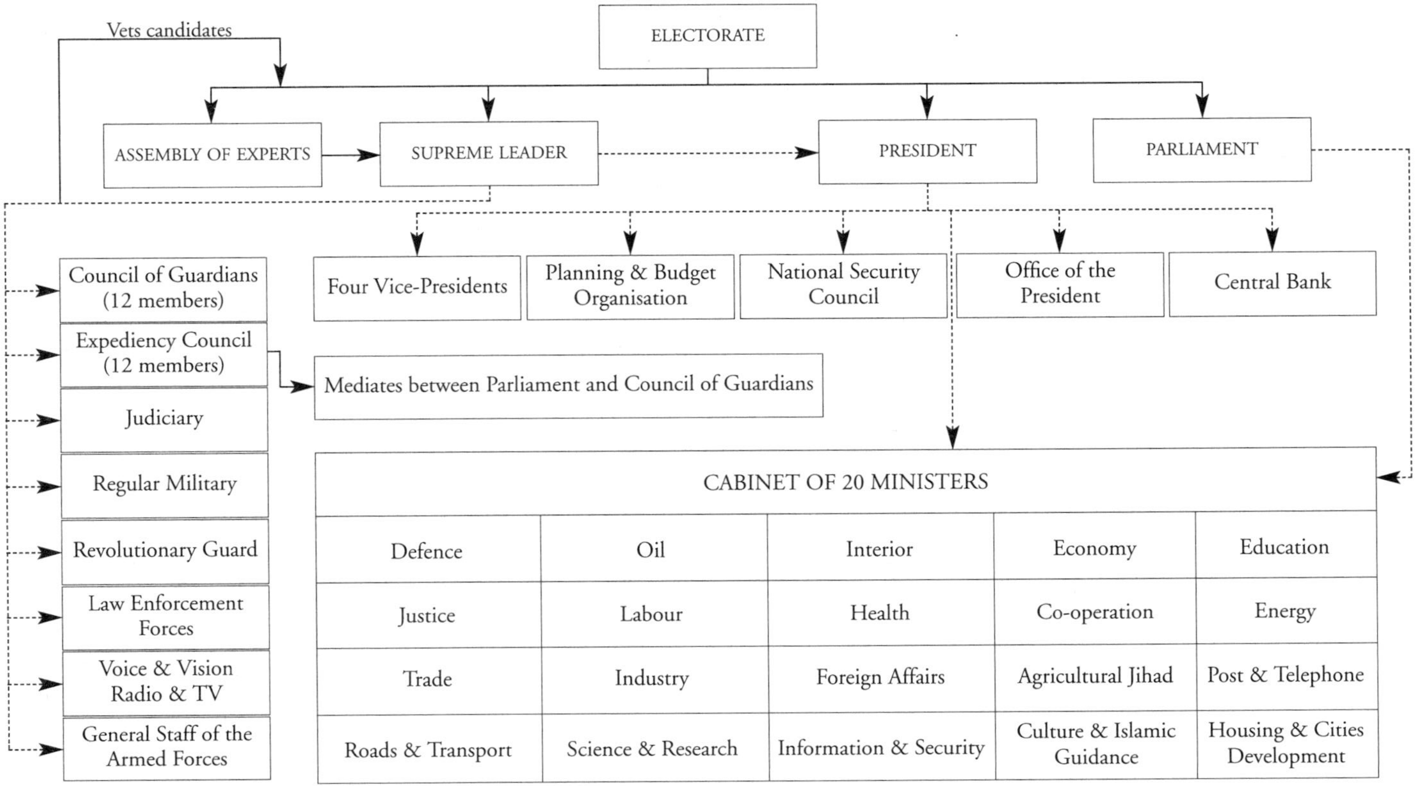

Key: ⟶ Elected ‑‑‑➤ Appointed/Confirmed
Source: The Economist Intelligence Unit (2006).

External Politics

The 1990s and early twenty-first century have been a time of increased international involvement in the political and economic affairs of Iran. Throughout this period, the same foreign policy issues have continued to plague the Iranian state. In parallel, Iran's distrust of, and tense relationship with, the West, especially the United States, have continued. UN sanctions have now been imposed and, though far from being decisive, they have sent a troubling signal to the political elite in Tehran that had counted on Russian and Chinese support. Furthermore, less visible US efforts to curb Iran's access to the international banking system has made foreign direct investment or the purchase of shares in privatised industries more precarious.

Beyond the background issue of the Iranian state's opposition to the West, three major foreign policy issues have arisen which have had a major influence on the Iranian privatisation policy. The first is the 1991 Gulf War and the second is the war on terror, which manifested itself in the invasions of Afghanistan and Iraq. Finally, the Iranian government's development of nuclear technology, especially the enrichment of uranium, has had a major influence on both political and economic policies of the country.

The Gulf War, 1991

Iran's foreign policy has been at times contradictory, in that it has not taken advantage of international circumstances (e.g. the 2003 war on Iraq) to pursue an aggressive policy towards long-standing enemies like Iraq (personal communication).[17] This is particularly interesting, since it tells us much about the Iranian political scene which had other values to uphold. Thus, during the 1991 war over Kuwait,[18] Iran limited its actions to condemning the United States, but otherwise remained neutral despite the fact that the war with Iraq had only recently ended. It refused to take part in the multinational invasion of Iraq and even allowed Iraqi aircraft to land on its territory for safe keeping. Moreover, Iraqi refugees crossed the border into Iran without let or hindrance. Such behaviour was exceptional in the region. Egypt, Syria, Saudi Arabia and other nations in the Middle East took part in the invasion. Unlike the more recent US-led war against Iraq, in 2003 joining the coalition would have had no adverse consequences for Iran. The Iranian attitude towards this particular event in 1991 points to the fact that solidarity with Islamic countries in general was paramount. The anti-American sentiment also played an important part in keeping Iran out of the fray.

Hezbollah

Iran's policy towards Israel took the form of a war of words and covert support for groups engaged in fighting or opposing Israeli actions both abroad or inside Israel. Support of Hezbollah, the Party of God, was an example of the latter.

The Hezbollah movement was formed around a group of Shi'ite clerics to fight Israel during its invasion of Lebanon in 1982. Before that, the weak Shi'as in Lebanon had coalesced around the Amal movement, a largely secular and relatively moderate group. In its initial stages Hezbollah drew the majority of its support from the more radical wing of the Amal group. It was based in the Bekaa Valley, southern Lebanon (site of the invasion) and southern Beirut. Apart from its motivation to drive out the Israelis, Hezbollah's long-term goal was to create an Islamic state in Lebanon. It was, and continued to be, a cross between a terrorist organisation, militia, guerrilla insurgency and quasi-government. It fought in Lebanon's 15-year civil war, which lasted from 1975 to 1990, as other militias did. At times it even fought against the Amal group. At the same time, it fought the Israelis until they withdrew in 2000. Hezbollah is also a service provider in Lebanon, offering many welfare services, traditionally supplied by the government, to the Lebanese people, including, for instance, medical care. More recently, it has emerged as a viable political party in Lebanon, simultaneously using terrorist tactics, including suicide bombings of various sorts. For this reason, Hezbollah was officially classified as a terrorist organisation by the United States after the 9/11 attacks (CNN, 2003).

From its inception, Hezbollah has received substantial support from Iran. Its animosity towards Israel made it a natural ally of the new post-revolution Iranian state. Iran has supplied Hezbollah with military equipment and also provided ideological sustenance, such support being channelled through the Revolutionary Guard. In effect, Hezbollah has been used as a front for Iran's policy towards Israel. Though Hezbollah has mostly been directed by Iran, there have been instances where the group has acted without Iran's consent. Another feature of Iran's policy has been to fuel the Israeli–Palestinian conflict (Globalsecurity, 2005).

Needless to say, such policies have sharpened the West's negative perception of Iran.

The war on terror and the "axis of evil"

Public sector reform and privatisation are public policy issues that are always embedded in other issues pertaining to structural adjustments, degrees of state economic intervention and the regulation of markets (Suleiman and Waterbury, 1990). It would be an exercise in futility to separate political issues – public resource allocation, the provision of collective goods and the distribution of wealth in society – from the larger policy context. Public sector reform and privatisation never take place in isolation, away from the broader framework of macro-economic and political adjustment.

Building on the negative perceptions already existing about Iran internationally and the belief that Iran was a state-sponsor of terrorism, President George W. Bush included Iran in the "axis of evil" (besides Iraq and North Korea) in his 2002 State of the Union speech. This designation has served as a reminder of the tense relationship between the United States and Iran. Whereas in the mid-1990s, there seemed to be a possibility for détente and even a revitalisation of US–Iranian relations, the "axis of evil" speech demonstrated that, at least from the American point of view, this would be impossible in the near future. The consequence of such hostility, along with the sanctions imposed on Iran by the United States, has been to drive away potential foreign investors in the badly needed development of the Iranian oil and gas industry.

The war on Iraq

The war on Iraq should be regarded within a broader context, that of the previous foreign policies of both Iran and the United States in the region, the former being that of non-interference and the latter that of maintaining a balance of power in the area. Iran continued its policy of non-interference in neighbouring Iraq as it had done in the 1991 Gulf War. However, the situation became more complex when the war in Iraq evolved into a state of insurgency and counter-insurgency (personal communication).[19]

Iraq is dominated by three separate communities; the Shi'a majority in the southern areas of the country, the Sunni minority in the middle areas and the Kurds in the north of the country. There is a significant degree of mixing throughout the country, in particular in and around Baghdad. The secular, socialist Baath party ruled Iraq from 1963 until its leader was ousted in 2003; the party and its leader, Saddam Hussein, were largely Sunni. From the Western point of view, the Sunni-dominated government under Saddam Hussein served as a convenient counterbalance to the Shiʿa-dominated Iranian government after the Islamic Revolution of 1979. Until the Iraqi invasion of Kuwait in 1990, the United States had backed Iraq, most conspicuously during the Iran–Iraq war (1980–88) when its support extended to include military assistance.

US foreign policy after 1991 was that of "dual containment" when America stopped playing one country off against the other and sought to limit their power and influence in the region – to "contain" them. The major instruments of this policy were economic and political sanctions. The sanctions imposed on Iran had been in existence since the Islamic Revolution and the storming of the US embassy; while the sanctions on Iraq were applied after the 1991 war to dampen Saddam Hussein's ability to produce weapons of mass destruction and also to punish him for denying the UN weapons inspectors full access.

The power balance between Iran and Iraq and, by extension, between Sunni and Shiʿa Muslims in the Gulf region, had been an important factor in Iranian–Western relations, but since the institution of a new democratically elected government in Iraq after the 2003 war, the balancing act has lost its gist. This is because the Shiʿa community constitutes the majority of the Iraqi population, a fact which is bound to determine the leadership of the country. The rising concern of some US-based analysts is that Shiʿa Iran may exert undue influence on a Shiʿa-dominated Iraq (personal communication).[20] Such an eventuality would threaten the regional balance of power and potentially have global ramifications. While such a situation is not visibly occurring at present, it cannot be discounted in the years ahead.

The balance of power between Iran and Iraq clearly has had an effect on Iranian foreign policy. Perceived or real threats from a large US military force in Iraq causes Iran to fear a possible attack on its territory. These fears are reinforced by certain instances such as the reported US reconnaissance missions by special forces. Another source of worry is that Iraq could become a failed state if a civil war were to occur there despite the recent lull in violence. For Iran, which already borders one failed state (Afghanistan), such a possibility poses an even greater threat. In either case, Iranian foreign policy is, and will continue to be, directly affected by changes in Iraq.

American paranoia over Iranian involvement in the Iraqi insurgency has triggered a more aggressive approach towards Iran (personal communication).[21] However, this matter cannot be divorced from the more prominent issue of Iran's nuclear programme.

Iran's nuclear programme

This section highlights the distractions which delay the privatisation programme in general and reforms in the oil and gas sector in particular. It argues that Iran's depleting oil reserves, the government's long neglect of proper maintenance of wells in the oil producing fields, growing domestic demand for fuel products, inadequacy of domestic investment and the lack of access to foreign credit and technology have resulted in a petroleum crisis in which oil exports (the lifeblood of the Iranian economy) are likely to vanish in less than a decade (Amuzegar, 2007).

While the Iranian nuclear programme has been going on for years, it has recently taken on increasing importance in world politics, largely due to America's preoccupation with the so-called "rogue states". The accusation that Iran's uranium enrichment programme is likely to lead to the development of nuclear weapons is directly linked to the bad reputation of Iran worldwide, especially since 1979, as an unreliable and potentially aggressive state. There have also been charges of Iran's intention to produce long-range missiles. The situation has led to extensive negotiations between Iran, on the one hand, and Britain, France and Germany, on the other.[22]

Iran's nuclear ambitions may have a considerable effect on the future economic situation. In particular, the negotiation process may lead to benefits for Iran if it plays its cards right. Specifically, the United States has dangled a carrot in the shape of membership of the World Trade Organisation (WTO) in exchange for abandoning its nuclear ambitions (personal communication).[23] Alternatively, increased economic sanctions and even an embargo could be imposed to ratchet up the pressure on Iran. In either of these two cases, the effects on the Iranian economy could be disastrous.

Iran's Economic Problems

Iran needs to revive its economy and address the acute problem of unemployment. High oil prices have provided the political leadership with revenues far above and beyond their expectations. This in turn has given it the luxury of providing some breathing space before the patience of those young unemployed and underemployed becomes exhausted. In the city council elections of mid-December 2006, where the turnout was a hefty 60% of the voters, the population at large showed its displeasure by voting against the President's (Ahmadinejad's) list of candidates. The President, who had been in office for three years, had failed to fulfil his campaign promise to put food on the table of every Iranian family from oil revenues.

Iran's economy depends critically on its oil and gas exports. According to a recent report released by the IEA in its World Energy Outlook through to 2030 (Clist, 2006), Iran will need to attract foreign investment and technology. The IEA estimates that it will need approximately $200 billion to enable its oil industry to maintain its current production rate through to 2030. With further sanctions looming, it seems quite hard to imagine that foreign investors would be willing to provide the necessary finance and/or technology to Iran to make any progress. In short, it is incumbent upon the current administration to take a less confrontational approach in its foreign policy so as to attract investment and avoid sanctions which could cripple the country's economy.

With the fast-growing domestic demand for, and slowly rising supply of, energy products, the Islamic Republic's populist policy of subsidisation of fuel consumption is adding further strain to the economy. On the demand side, by paying the highest energy subsidies among the 22 countries in the Middle East and North Africa, the government has artificially raised consumer demand for various energy products far beyond international norms.[24] With gasoline and other fuels sold at a fraction of their world averages, total yearly energy subsidies in Iran, according to a major newspaper report based on the oil ministry's figures, range between $30 billion and $42 billion a year (Amuzegar, 2007), or 16–23% of GDP, depending on f.o.b. prices in Iran's southern ports, or the prevailing consumer prices in Europe. As a result, Iran's appetite for the highly subsidised fuel products has arguably become the world's highest (with annual consumption of electric power, natural gas and gasoline growing by 7%, 9% and 12% respectively), while its population is growing by about 1.55% a year, and real per capita income by less than 4%. With most fuel prices at about one tenth of those in neighbouring countries, an estimated $1 billion worth of energy products are annually smuggled out of the country (ibid.).

Economics and Islam

Since the 1979 Islamic Revolution the Iranian economy has been in a state of continuous crisis. This has been largely due to conflicting political and legal interpretations of the ideological objectives of an Islamic economy, as well as to the nature and functions of the state in an Islamic society (Shahriari-Rad, 2002). Economic manifestations of this crisis are insecurity about private capital ownership and property rights, one that has detrimental effects on the process of production in general, causing shortages of basic commodities, hyperinflation and stagnation of the economy. This situation has been further aggravated by the accumulation of foreign debt, a substantial increase in "rent-seeking" activities at the expense of productive enterprise, a shrinking of capital stock, an inadequately trained workforce and a lack of reform in economic institutions (Khavand, 1999).

The direction of the Iranian economy is still influenced by religious debates. More specifically, the privatisation policy is the result of negotiation between economic development, social justice and cultural identity. In all four periods, Iran's rulers have largely failed to overcome the conflicting demands of economic growth, social justice and a conservative, Islamic cultural environment. The Shah embraced economic growth at the expense of other objectives, alienating almost the entire population in the process (personal communication).[25] The various leaders of the Islamic Revolution have, at different times, promoted economic growth and social justice in the context of an Islamic government, but have been stymied by conflicts within governing coalitions over the appropriate balance between the two aims and their conformity with Islamic doctrine. Certainly, Iran is not alone in confronting these kinds of tensions. Other societies, not least those of the West, face similar challenges, and at different times have attempted a variety of political strategies to overcome them. Generally, where societies have been perceived to succeed in this regard, it is by promoting one or possibly two objectives at the expense of the other(s).

This must lead one to pose the question: are economic growth, social justice and the maintenance of the cultural status quo an "impossible triple"? Rahnema and Behdad (1995) argue that the attempt to implement Islam as a political philosophy is bound to disappoint because it promises to achieve all three objectives, despite the fact that laissez-faire economics (presumably conducive to growth) and social justice are in fact inimical:

> While Islam promises social justice to the poor and destitute, it also makes it clear that the quest for worldly possessions, the accumulation of property and profit, is not only accepted but highly respected . . . The Islamic revivalist movements promise the establishment of a social order that will unite conflicting aims: the welfare of the poor and the security of property (Rahnema and Behdad, 1995:12).

The authors go on to argue that, under the conservative–pragmatist coalition associated with Rafsanjani and Khamenei, social justice was downplayed in an attempt to attain cultural and economic imperatives. Khomeini's rhetoric (in the manner of Abu Dharr's), to liberate the *mostazafin* (oppressed) from the *mostakberin* (the high and mighty or 'fat cats'), was dropped.[26] Abrahamian concurs, arguing that Rafsanjani and Khamenei, since Khomeini's death, have continued denouncing imperialism, Zionism and "exploitative capitalism" while, at the same time, protecting petty-bourgeois property and further toning down their radical rhetoric. They no longer talk of land reform, income redistribution or nationalisation of foreign trade (Abrahamian, 1991:119).

The Khatami era, which saw a more liberal approach to social policy resulting from a rapprochement between the radicals (leftists) and the moderates, now seems something of an aberration. As Wells argues:

> It is clearly the revisionist right that has the upper hand; so long as they control the courts and the faqihship, hard-liners and moderate reformers will be on the defensive . . . Thermidor may be slowly and painfully coming to a close with Islamic Iran exiting its era of leftist supremacy only to embark on a new phase of revisionist right-wing domination (Wells, 1999:39).

This view was confirmed by the victory of the conservative candidates in the 2004 elections to the Majlis. If the Islamic Republic has largely failed to master the contradiction between development, equity and cultural imperatives, it has, to an extent, allowed Iran to move away from the wholly "arbitrary state" identified by Katouzian (2000; 2003) as the norm for Iranian political institutions. Whether this is a long-term development is open to question, as Khatami's project of "Islamic democracy", based on strengthened civil institutions, appears to be one of the main casualties of the conservatives' revival.[27]

Iran witnessed a further intensification of the Republic's chronic factionalism. Recent sabre-rattling over the country's nuclear facilities – in open defiance of the "Great Satan" (the United States) and other Western powers – suggests that the clerical establishment may also be engineering a return to Khomeini's policy, or raising the spectre of foreign intervention and domination to mobilise support internally. The anti-Western discourse, which previous Iranian leaders have used in the past, does not seem to help the Iranian economy. Khatami realised this and tried to promote an openness to foreigners. An example of this policy was reflected in the government's rapid acceptance of foreign aid after the earthquake of 2004 in Bam, which stood in contrast to the rejection of similar offers that followed Iran's previous big quake, in 1990. Even the Council of Guardians was persuaded to ratify legislation designed to attract foreign investment (Anon, 2004b). Revisionist or conservative domination is unlikely to derail Khatami's economic reform programme, mainly because the conservative coalition is at least as committed to privatisation and (economic) liberalisation as was the broad leftist coalition associated with the President. During the 2004 election campaign, conservatives stressed the need to eliminate factional struggles and concentrate on economic reform to attract inward investment (Muir, 2004). The incoming President Ahmadinejad seemed to be in favour of encouraging foreign and domestic investment, but still used a cautious, or rather nationalistic, tone when talking up the role of Iranian nationals living abroad rather more so than the role of alien or Western companies. He said:

> One of the main topics of our economic policy is the expansion of foreign and domestic investment . . . We will especially use our dear Iranians [currently living abroad] who are ready to take part in developing their country (BBC, 2005d).

Whether this is attainable, given Iran's recent history of chronic political factionalism, the continuing friction between state power and the wider society, and the tension caused by Iran's strained foreign relations, is more open to question.

Economic development

The Iranian economy is in rather a bad shape. Iran's population has doubled since the revolution, dramatically increasing consumption. With little growth in the output of the economy (GDP), the standard of living has nearly halved. Half the population, including 70% of government employees, live below the official poverty line. Many have more than one job just to make ends meet. Unemployment is estimated to be 25%, and the future seems bleak. Every year 800,000 young people enter the job market. It is estimated that more than $100 billion needs to be invested to maintain current employment levels, but only $2.3 billion is currently allocated to job creation.

After 1989, the critical division was on economic policy, between the radicals (Maktabis) who clung on to Khomeini's early populism, and the conservatives whose economic views reflected the laissez-faire approach of their bazaar constituency. The balance of power between the two factions ebbed and flowed, whilst prominent power brokers – notably President Rafsanjani and spiritual leader Khamenei – attempted to implement a pragmatic reformist agenda with little success. As Owen argues, the chronic factionalism of the period (1989–97) resulted in "a stand-off between the two main factions in which few new policy initiatives were possible at the national level" (2000:112).

The conservatives' improved showing in the 1992 Majlis elections, which left them in control of most of the institutions of state and with an ally (on economic matters particularly) in President Rafsanjani, should have cleared the way for economic reform. However, the right was hamstrung by its "reluctance to compromise on a number of critical issues, most of which revolved around foreign and cultural policy" (Wells, 1999:33). Having attained power, the conservatives undermined their coalition by weeding out socially liberal clerics and technocrats. This inability to forge a consistent position on economic development and cultural/political issues impaired the conservatives' ability to implement their economic reform agenda.

President Khatami was fully committed to restructuring the economy. His government approved the privatisation of 538 state-owned companies through auctions and stock sales, with plans to privatise another 2,000 (Reuters, 1999). The Iranian parliament (Majlis) approved plans to overhaul the oil industry and boost Iran's sagging oil production from 3.5 million bpd (barrels per day) to 6 million bpd over the next ten years (USEIA, 1998). This required at least $120 billion in capital investment, which had to come from private investors and foreign sources.

One major factor affecting the development of the Iranian economy is the effect of subsidies. These amounted to over $7 billion a year for foodstuffs and more so for petrol. They have led to chronic budgetary problems for the Iranian government. In 1998, fuel subsidies alone reached $11 billion per year – double the country's development budget. Food subsidies for basic commodities such as bread, rice and sugar accounted for another $2.2 billion per year. Together, these subsidies used up all of the country's $16 billion oil income for the year (Reuters, 1998b). President Khatami came to accept market reform and, with no other source of capital available to the country, advocated attracting foreign investments. He skilfully combined cuts in subsidies, privatisation of state firms and changes in labour laws – in short, he applied shock therapy to the market – with calls for the liberalisation of social and political rules to allow freedom of expression, freedom of the press and the establishment of a civil society. Most recently, the burdensome costs of subsidies for public finances has forced parliament to present a draft law that aims to cut down on the amount of subsidised petrol that motorists can consume. One should mention that what is beneficial for the economy as a whole may not necessarily elicit support from the population. This last measure, if implemented, would certainly contradict Ahmadinejad's electoral campaign promises and therefore undermine his popularity (Coville, 2005).

Economic reform since 1989

Briefly, the evolution of the Iranian economy since 1989 has been one of economic reform of varying degrees of intensity. Broadly speaking, there have been two major ways by which this has been occurring. The first has been attempts to diversify the oil-dependent economy into others areas. An example has been a growth of investment in the petrochemical industry from oil revenues (PIIC, 2000). Through a strategic development plan and the implementation of various petrochemical projects, efforts are under way to raise the production level of Iran's petrochemical industry. Begun in 1997, the plan is divided into five phases, lasting until 2013. By the end of this period, the total volume of final products is estimated to be 16.8 million metric tons per year. The value of investment will have amounted to $20.6 billion, with total sales topping $11.8 billion per annum (ibid.).

The services sector saw the fastest long-term growth in terms of its share of GDP, but currency exchange restrictions, excessive bureaucracy and the uncertainty of long-term planning made this a volatile sector. In the agricultural sector, state investment boosted output, while the liberalisation of production and the improvement in packaging and marketing helped to develop new export markets. Large-scale irrigation schemes, together with the wider production of export-based agricultural produce, such as dates, flowers and pistachios, engendered the fastest growth of any sector in Iran over much of the 1990s, although successive years of severe drought in 1999, 2000 and 2001 held back output growth substantially. Agriculture was one of the largest employers, accounting for 22% of all jobs, according to the 1991 census, and remains so today (Economist Intelligence Unit, 2004). The other method whereby economic reform was initiated, consisted of attempts to attract foreign investment by creating a more favourable climate for potential investors such as reducing restrictions on the importation of capital goods and lowering import duties. In addition, free trade zones were established.

The financial sector

The financial sector in Iran was transformed following the 1979 Islamic Revolution with the nationalisation of all banks and the revamping of the financial sector according to Islamic principles which were enshrined in the Law on Usury-Free Banking of 1983. In recent years, the government has implemented reforms in the financial sector designed to conform more closely to international practices. However, heavy government intervention has remained unchanged (*Iran Daily*, 2004).

The banking sector in Iran is run according to Islamic law, which prohibits interest payments. Only since 2001 have private banks been authorised to operate and foreign banks been given permission to operate in Iran's free trade zones. Currently, the banking system consists of six state-owned commercial banks, four state-owned specialised banks, a state-owned postal-operated bank (licensed in 2004) and four recently established small private banks.[28] State-owned commercial and specialised banks dominate the banking system, holding about 98% of deposits. Iran's financial system also includes a number of small banking institutions and informal financial intermediaries. About 6,000 *Qarz-ul-Hasanah* funds, which raise zero-interest funds and provide interest-free loans, operate on a small scale. The sector is supervised by the Bank Markazi, the central bank. The World Bank and the IMF have been providing assistance to enhance banking supervision by the Bank Markazi.

Iran's insurance sector is still dominated by the state, but has begun to move along the path to reform and liberalisation. Following the revolution, all insurance companies were nationalised and the sector still remains under state control, with approximately 50% of policies issued for compulsory third-part liability car insurance (Dana, 2004). Recent reforms, however, have allowed private insurance companies to operate in the free trade zones. There are currently five major state-owned companies, five private domestic companies and one foreign representative office, the first of its kind, opened by a UK firm in Tehran in 2004. The Central Insurance Authority (Bimeh Markazi) is the regulator and provides reinsurance.

The Tehran Stock Exchange (TSE) reopened in 1989 and has performed well in recent years, with over 300 listed companies, the majority of which are parastatal enterprises with varying degrees of direct and indirect government ownership. The TSE has also started a modernisation programme and has opened regional branches in other cities in Iran. There is no independent supervisory entity that oversees issuance and the trading of securities. Before his election as President, Ahmadinejad denounced the Tehran Stock Exchange, equating it with gambling. Although his position quickly changed once he took office, the market nonetheless took a serious blow. According to a survey by Karafarin Bank,[29] the Tehran Stock Exchange reached its lowest point nine months into Ahmadinejad's presidency. The index registered a 26% decrease compared to its peak at the end of the first month of the year. The drop affected financial stocks by 44% and industrial stocks by 20%.

Oil

Oil continues to play an extremely important role in the Iranian economy. Iran is still the second largest oil producer in OPEC (Organisation of the Petroleum Exporting Countries). It holds 10% of the world's proven oil reserves. In addition, what may become even more important in the future is that Iran holds the world's second largest natural gas reserves (the largest belonging to Russia) (personal communication).[30] Clearly, such large oil reserves are bound to dominate the Iranian economy. The disadvantage of a high degree of dependence on one commodity, in this case oil, is the fluctuations in world prices. In fact, it is possible to trace Iran's internal economic performance in terms of changes in world oil prices. For instance, the strong oil market in 1996 led to increased government revenue for Iran which eased financial pressures on the country and allowed for timely debt repayments. Then in 1997, the fall in world oil prices led to a tightening of public finance. Again, in 1999/2000 the rise in world oil prices reversed the trend, boosting government funds. Such fluctuations in the world oil prices can be traced to the present day and is bound to continue to affect not only the economic but also the political situation in Iran.

In particular, we may predict that rising oil prices will have a dampening effect, other things being equal, on the drive for economic reform in the country (personal communication).[31] The reasoning is that with higher oil prices the government has more money in its budget to continue to pay the subsidies mentioned above. These provide a disincentive to want to change the political status quo – that of an economy driven by oil and subsidies. Thus, high oil prices set in motion a self-reinforcing cycle. It is when oil prices drop that we may expect (other things being equal) calls for reform of the economy to increase. With lower oil prices, the government is left with less money to fund the burdensome subsidies. This in turn may translate into civil unrest which may trigger demands for reform. Thus, low oil prices create a self-reinforcing cycle. Fluctuations in prices, therefore, show clearly the sort of interaction of politics and economics that help determine changes in the Iranian economy.

Investment is badly needed in the oil industry as this would benefit the entire economy. A recent IMF report on the Iranian economy predicted that Iran needed to mobilise $4 billion a year in foreign loans and direct investment if it is to achieve an average growth rate of 5.4% per year to stabilise unemployment (IMF, 2003:34). The report assumes that Iran could secure significant foreign investment in the oil industry and that it would be able to borrow $3 billion a year. If Iran did not have access to that kind of foreign capital, it would suffer directly from a shortfall of $4 billion a year and, additionally, would suffer further from the loss of the extra oil exports that those funds would finance. In particular, the IMF forecasts that foreign funds would allow Iran to expand its oil exports from 2.3 million bpd in 2002/03 to 3.9 million in 2008/09. However, if oil exports stagnated at the 2002/03 level, Iran would run a deficit of $11 billion a year by 2008/09. In other words, foreign funds together with the expansion of oil exports are central to Iran's economic plans and could be key to preventing social unrest (Clawson, 2003; IMF, 2003:34).

The country has failed to properly maintain its oil installations since the revolution, particularly after the severe damage they had sustained during the Iran–Iraq war. As a result, the recovery factor in Iran's oil fields has declined to a range considerably below that of international norms. Yet, while some oil deposits might have been permanently lost as a result of poor maintenance, oil-well capacity has not been irretrievably damaged and the ongoing gas injection programme has significantly raised the recovery factor (Amuzegar, 2007). Accordingly, official data show that since the end of the Iran–Iraq war in 1988, Iran has actually increased its oil reserves more than it has taken out of the ground.[32] In addition, due to some one million b/d condensates from the South Pars gas fields being added to the daily output, an estimate formed by the *Middle East Economic Survey* (MEES) shows an actual rise in Iran's domestic production in recent years (Amuzegar, 2007). In short, based on official[33] and generally accepted reserves figures, Iran will, at the current rate of extraction, produce and export oil for the next 70 years.

To sum up, Khomeini's successors began to introduce reforms in economic policies, though repression of personal freedoms continued unabated. Even so, the people boldly risked the wrath of the Revolutionary Guards by installing (camouflaged) satellite dishes to tune in to news broadcasts from around the world. The use of computers and the internet swept across society, a phenomenon that further helped the masses to learn more about international relations, and became more politically active as a result.

A more liberal climate emerged with the ascendancy of the modern cleric, Mohammad Khatami, who was elected president in 1997 and again in 2001, bringing hope and optimism to the non-conservative communities in the country. His liberalising policies at home were matched by moderation in foreign affairs, showing a willingness to open a dialogue with the United States. His overture, however, was quickly quashed by the less pliable Supreme Leader Khamanei. Similarly, Khatami's social reforms were frustrated at every turn by the conservative elements in the clerical establishment.

With the election of Ahmadinejad to the presidency in 2005, Khatami's liberalism was checked. Though Ahmadinejad was elected on his promise to help the poor and stamp out corruption, he did not deliver once in power. In the economic field, Khatami's reforming initiatives endured, albeit precariously. But needs must, and reforms had to move forward, with privatisation gradually spreading, despite the endemic obstacles. The much needed foreign investments in the all-important, but aging, oil and gas industry was not forthcoming as it was hoped, partly because of the mishandling of foreign relations. Huge subsidies remained undiminished, straining the public purse.

How far Iran will go down the path of liberalisation and reform is a question that is addressed in the next two chapters.

NOTES

1 During the Second Republic, the Expediency Council increased from 13 to 20 members under the 1989 Constitution. It was chaired by the President and members chosen by the *faqih*. "In its current form, thus, the control of the council remains with the 'Leader–President' coalition. Despite the presence of Ahmad Khomeini, former Majlis Speaker Mehdi Karrubi, former PM Moussavi and former Prosecutor-General Mohammad Khoini'a as the Old Guard hard-liners and their counterweights in Rafsanjani and Ayatollah Yazdi and Mahdavi Kani in the council, it is clear that the 'institutional' appointments ensure a bias towards the realists' (or pragmatists') line" (Ehteshami, 1995:47). Constitutionally, besides the six Islamic jurisprudent members, the heads of the three branches of government and the Minister of the Interior, the Cabinet Minister and Head of the Majlis Committee concerned with the matter under discussion were always present. The clerical establishment continued. This was estimated to comprise between 90,000–200,000 members and around 200 Ayatollahs. "[Clerics] continued to form the backbone of post-Khomeini Islamic Republican Iran" (ibid.).

2 More than 40% of the country's 65 million citizens are under the age of 15.

3 He served in the Central Council of the Party, also as Deputy Ministry of Defence, representative of the Council in the ministry and Commander of the Islamic Revolutionary Guards Corps. For more details see Iranchamber (2005).

4 In 1981, Hojatoleslam Khamenei was elected President of the Islamic Republic with 95% of the votes cast in his favour.

5 Rafsanjani undoubtedly qualifies as a pragmatic careerist.

6 Further evidence of this view is provided by Wright, who points out that Rafsanjani's daughter, Faezeh Hashemi, was the publisher of *Zan*, a woman's magazine which took up a campaigning position on social issues until it was banned in 1999 (Wright, 2000).

7 Personal communication. Discussions with Dr Bijan Bidabadi, Faculty Member & Director of Foreign Exchange Research Department of Central Bank of Iran, Tehran, December 2005.

8 As demonstrated by the moderates' subsequent control of the Majlis (2000–04), the critical levers of power were in the conservative hands of the Council of Guardians, the Expediency Council, the *faqih* establishment and the judiciary. Iran's (1989) constitution, which granted veto power to the Council of Guardians and gave the Expediency Council the key role of adjudicating between parliament and the Council of Guardians, effectively sidelined parliament. Hence Milani's characterisation of the Islamic Republic as a regime based on "*limited* popular sovereignty" [my italics] (Milani, 1997:82).

9 Ahmadinejad also drew widespread criticism from the West when he told a conference in Tehran that the state of Israel should be "wiped off the map" (BBC, 2006).

10 The Expediency Council mediates disputes between the Council of Guardians and the Majlis, and tends to rule on the side of the former. Rafsanjani took over as Chairman of the Expediency Council when his second Presidential term ended in 1997, and he has used the post to ensure that he continues to command influence at the heart of the Islamic Republic. The Assembly of Experts is an elected all-clerical body with the primary task of selecting the supreme leader and the members of the Council of Guardians. The Assembly can also theoretically dismiss the supreme leader if he fails to meet specific criteria or becomes unable to execute his duties satisfactorily (Economist Intelligence Unit, 2005).

11 Personal communication. Discussion with Dr Narsi Ghorban, Managing Director, NarKangan (NGTLC) Gas to Liquid International Co., Tehran, December 2005.

12 Personal communication. Discussion with Dr Narsi Ghorban, Managing Director, NarKangan (NGTLC) Gas to Liquid International Co., Tehran, December 2005.

13 Personal communication. Discussion with Dr Narsi Ghorban, Managing Director, NarKangan (NGTLC) Gas to Liquid International Co., Tehran, December 2005.

14 Since the 1990s, the left's main political vehicle has been the Society of Combatant Clergy of Tehran (Wells, 1999).

15 The *Ruhaniyat*'s full title is *Jameh-e Ruhaniyat-e Mobarez-e Tehran* (ibid.).

16 Khatami had been forced out of his former ministerial position by the conservatives in 1992 (ibid).

17 Personal communication. Discussions with Dr Mehdi M. Behkish, Secretary-General, International Chamber of Commerce-Iran, and Associate Professor in Economics, December 2005.

18 Iraq invaded Kuwait in August 1990. Saddam Hussein had at least a threefold plan for the invasion: to obtain Kuwait's large oil reserves, cancel Iraq's existing debt with Kuwait, and gain a stronger regional position in the balance of power. Immediately after the invasion, the United Nations condemned it and ordered Iraq's withdrawal. In order to protect Saudi Arabia from the possibility of further aggression, the United States and other nations sent forces to Saudi Arabia as a deterrent force. The United Nations Security Council authorised the use of force against Iraq in November 1990. By January 1991, the international forces had grown to around 700,000 troops. On 16 January 1991, the US-led invasion of Kuwait, known as Operation Desert Storm, began with US air strikes. Hostilities were over by 28 February 1991, when the United States called a ceasefire. By this point, the Iraqi military had completely collapsed.

19 Personal communication. Discussion with Dr Jafar Ghamat, lecturer, Tehran Azad University, December 2004.

20 Personal communication. Discussions with Dr Mustafa Zahrani, former Director of the Institute for Political and International Studies, December 2004.
21 Personal communication. Discussions with Dr Mustafa Zahrani, former Director of the Institute for Political and International Studies, December 2004.
22 It is not clear what the solution to the problem is. As one analyst puts it: "Western options range from the poor to the abysmal. Military approaches are relatively easy to dismiss. Iran is simply too big to invade, and the United States is militarily overstretched and diplomatically unpopular even with its traditional allies because of the war in Iraq. A narrow strike on Iranian nuclear facilities similar to what Israel did in Iraq in 1981, when it destroyed the Osirak nuclear reactor and set back Iraq's bomb, would probably fail, as Iran has dispersed its facilities" (Byman, 2005).
23 Personal communication. Discussions with Seyed Hossein Mousavian, former Speaker of the Supreme National Security Council, Tehran, November 2004.
24 Personal communication. Discussion with Mr M. A. Ala, March 2007, London.
25 Personal communication. Discussion with Mr Hossein Nasiri, former Adviser to President Khatami, Tehran, July 2004.
26 Abu Dharr was a companion of the Prophet Mohammad associated with defending the cause of the *mostazafin*. He was exiled by the third Caliph for his excessive radicalism. His version of Islam is associated with radical interpretations of the religion, arguing for social justice, land reform and the like; see Rahnema and Behdad (1995:12).
27 Other observers tend to be more optimistic about the future of reform in Iran: "The ebbing of political Islam is bringing about a detachment from religion. The fact that the revolution took place means that the flow can run only in the direction of secularization" (Roy, 1994:181).
28 Pronouncement of Article 44 of the Constitutional Law in June 2004.
29 Personal communication. Discussion with Dr P. Aghili, Managing Director of Karafarin Bank, Tehran, December 2006.
30 Personal communication. Discussion with Dr Mohammad R. Arasti, Assistant Professor, Sharif University of Technology, Tehran, December 2005.
31 Personal communication. Discussion with Engineer Mohamed Hossein Nejad Hosseinian, Deputy Minister of Petroleum, International Affairs, Tehran, December 2005.
32 Personal communication. Discussion with Engineer Mohamed Hossein Nejad Hosseinian, Deputy Minister of Petroleum, International Affairs, January 2007.
33 Ibid.

5
PRIVATISATION IN IRAN

Introduction

This chapter directly addresses the specific determinants of Iranian privatisation policy and focuses on the four primary factors which directly affect privatisation policy, namely: a) the specific background of Iranian thinking on privatisation; b) the legal and institutional framework; c) the nature of oil and gas in relation to privatisation in Iran; and d) the personal thoughts of the policymakers concerned.

The account of the legal and institutional framework is based on a reading of the relevant legal documents concerning privatisation post-1990 in Iran. It also explores the institutional issues which have arisen since that time and which have constrained policy decisions. As the oil and gas sector is unique in Iran the contemporary determinants of its privatisation need to be addressed separately from other types of privatisation.

The personal views of decisionmakers who are concerned with privatisation were garnered through extensive interviews and a questionnaire. The questions put to the interviewees are explored here so as to provide a perspective on what kind of information is relevant and what sort of bias can be expected. The results of the interviews are then described and examined.

Background

In order to gauge the extent to which privatisation relates to economic development, the latter concept needs to be seen in the particular context of the country in question. In the case of the Islamic Republic of Iran, religious cannons and cultural values are obviously a factor to be considered on a par with the specific political and social setting. Modernisation and growth along Western lines of organisation and management – themselves based on private ownership and *laissez-faire* policies – need to be considered alongside the firmly established social traditions and beliefs (Shahriari-Rad, 2002).[1] The Qur'an itself recognises that growth and change should come naturally, as they do through the cycles of human life. Hence, the necessity for combining "effective" with "appropriate" economic policies and development strategies (Poku and Pentiford, 1998).

No privatisation plan should dare go against or challenge the legacy of the past and the deep-rooted cultures and values in society, otherwise its implementation would take place in an overtly hostile environment. Being a crucial means for change, privatisation must serve a legitimate social purpose and meet pressing human needs. This way, it is less likely to give rise to populist political discourses that would ridicule, or even cancel, its achievements. In addition, it must take into account what is tolerable within a given society and how far ownership transferral can go before generating a debilitating backlash. Privatisation policies thus need to be carefully crafted with a view to ensuring their proper implementation by taking on board as many influential constituents as possible.

At present, Iran seems to be committed to some level of privatisation, both in terms of action and political discourse. This is a subject about which interpretations of Islamic doctrine diverge. The different interpretations range from a more conservative approach to capitalism, one that views the accumulation of wealth with suspicion and strives to limit its magnitude as well as its employment, to a more liberal outlook, one which calls for some degree of synthesis between Islamic principles and modern thought (Zubaida, 1993).

Despite these divergences, a consensus has been reached based on the observation that the economies which adhered to the principles of Islam were neither purely capitalist nor avowedly socialist. Instead, they were seen as following a "third path" of economic management by the state, one where both the material and the spiritual needs of the people were taken into account. Under this ideology, private property is not recognised as absolute; it is subject to the limitations defined by *sharia* rules. Since the ultimate owner is God, the holders of property rights are accountable to him for how they exercise them. The state is a necessary arbiter in this relationship. In this sense, the Iranian economy is a particular mixture of private and public ownership.[2]

Capitalism is rejected on the grounds that it produces great disparities of income and property distribution. Its emphasis on individual self-interest and the unfettered pursuit of profit in business is seen as undermining social cohesion and depriving societies of their humane nature, the sense of brotherhood and a spirit of cooperation. The (almost) absolute endorsement of property rights, individual ownership and free enterprise result in antagonistic competitive practices that are detrimental to the socially just distribution of resources. Seen from the viewpoint of Islamic political economy, capitalism is an inherently contradictory economic philosophy where greater production and growth do not necessarily translate into higher levels of welfare for the majority of the people (Sadr, 1989; Abrahamian, 1993).

Similarly, socialism is rejected on the grounds that it limits the chances of personal development because it curtails freedom and choice, and bestows on the state a dominant economic role. Moreover, class struggle is considered to be inimical to the Islamic way of life, where a harmonious society is the result of all individuals working together for the good of the whole society. The corruption which inevitably accompanies ubiquitous economic ownership and management by the state has also been criticised, as is the employment of surplus resources for financing the internal security apparatus for the repression of dissenting opinions. Taking on board some positive aspects of socialism, such as central planning, the provision of welfare to the people and the ability to reduce unemployment, Islamic scholars have redefined the desirability of these features with exclusive reference to the *sharia* (Mannan, 1989).

Historically, the presence of aspects of the two (largely) opposing positions (capitalism and socialism) in Islamic economic thought has resulted in conflicting strategies of economic management undertaken by the Iranian state since the Second World War. For decades, at the beginning of the twentieth century, Iran's economy remained fractured, largely because of the control wielded by the Anglo-Iranian Oil Company over the oil industry. The company was an independent economic enclave with no forward or backward linkage effects on the rest of the economy (Nazmi, 1989:33–54). The private sector was small and insignificant, while the state did not command sufficient resources to build a strong base for state-owned industrial enterprises to generate economic growth and secure reinvestment funding.

With this objective in mind, but also guided by the desire to remedy a historical injustice, the National Front majority in the Iranian parliament passed a nationalisation law in 1951 (personal communication).[3] Under it, all of the oil industry that was developed in the preceding decades by the Anglo-Iranian Oil Company entered the portfolio of the state. The latter thus acquired full property rights, including controlling rights, exploitation rights and the right to the profits from the sale of oil, which earlier had been channelled to Britain. Oil export sanctions were the logical answer of the British government, one which prompted Iran to adopt a policy of import substitution. This resulted in an increase in domestic production and a modest industrial development and helped the national bourgeoisie to flourish. Some of it was financed by the state itself, which raised revenues by selling its gold reserves held in Canada and South Africa.

In the early 1950s, there was a brief involvement by the World Bank to settle the dispute between Iran and Britain by proposing that the oil revenues be distributed between them. The National Front leader, Mossadeq, who was then Prime Minister of the country, rejected the plan as it offered a price for Iranian oil significantly lower than the current market price. When Mossadeq requested that America provide additional aid in times of high inflation and was rebuffed, the Iranian economy deteriorated for the lack of badly needed rehabilitation cash. In these conditions a conflict arose between Mossadeq, supported by the bazaars and the national bourgeoisie on the one hand, and the Shah, supported by the landed gentry, on the other. At this point the religious community was divided in its allegiance (being the largest landowner in Iran), while supporting the nationalisation of the oil industry.

When oil production and export resumed, the economy quickly picked up speed. Between 1953 and 1978, under the Shah's rule, the Iranian economy grew at a rate matched by only a few other countries. Nazmi refers to the 1960s as a relevant example, during which time the real GNP increased at an annual rate of 7.5%, and also the years between 1965 and 1973 when the real GDP increased by an annual average of 10.4% (Nazmi, 1989:40). This process of economic growth, although beneficial in terms of social development (reduced infant mortality rate, increased life expectancy, etc.), paradoxically brought about conditions for overall underdevelopment in Iran. The main reason for this was the excessive reliance on economic rent from natural resources, one which had shaped class relations and provided an incentive for economic agents to co-exist passively alongside the national oil industry.

In such conditions, privatisation took the form of a state-led development of a rudimentary private sector within a mixed ownership regime (personal communication).[4] Before the Islamic Revolution, the emerging entrepreneurial class in Iran relied exclusively and unequivocally on resources supplied to it by the state apparatus. Thus the state, instead of being an impartial regulator of a vibrant private sector, became its most important and influential sponsor and, by implication, its unwitting controller. The state could afford to help the national entrepreneurial class, not only by direct financial means, but also by passing legislation that protected it from foreign competition, such as tax cuts and subsidies.

According to Nazmi (1989), the private capitalist class in Iran did not come about indigenously by individual entrepreneurs plying their skills and investing and reinvesting profits in directly productive activities, but by the intervention of the state. This was so because the demand for oil was such that the state could collect considerable economic rent and artificially finance a certain layer of society. Hence, the tendency of this class to regard the state as a benefactor and mentor. It was not born out of the development of the internal relations of production and could not be an independent economic force in and of itself. Thus, the private sector in Iran was doomed to consist of people who, often with relatively little capital of their own, acted as agents of the state empowered to use public funds.

In the 1960s, the Shah with large financial resources at his disposal, and aiming to energise this emerging entrepreneurial class, started to pursue policies of economic and social modernisation. At the same time, industrialisation of the country was supported by the West. The Iranian economy thus became a workable system which, even if experiencing certain problems, was able to adjust and survive in the face of the economic challenges of a globalised world (Rubin, 1981). It has to be acknowledged that the Pahlavi dynasty managed to generate huge cashflows and attract foreign investors into Iran. The Shah created opportunities for economic modernisation and development by rapidly expanding the basic infrastructure of the country. However, as Shahriari-Rad (2002) argues, these efforts proved unsustainable and cut across both the socio-cultural and economic capacities of the country to absorb change.

It was this discrepancy that gave strength to a rising discontent, fuelled by the religious clergy and supported by the bazaars. The Islamic Revolution of the late 1970s emphasised traditionalism in economic development, social justice and sovereignty. It was opposed to foreign influence over the Iranian economy and society, as well as to the local actors who promoted it (Khalili and Manafi, 2000). In a way, the revolutionary mood in the country became self-reinforcing. The rapid economic change and undue foreign presence generated an open opposition. Once it became a palpably potent social force, it caused a sense of insecurity, prompting insiders to embark on an asset-stripping spree of banks and private enterprises. This in turn made the need for a revolution even more pressing as it reinforced the case for it with convincing arguments. In the years following the revolution, the private sector was largely stifled, and economic initiatives returned to the state which launched a five-year plan.

Having lived through more than two decades of Islamic revolutionary rule, Iranian society and the body politic now seem to be decisively turning to the modernisation paradigm again. An important part of this is privatisation and the drawing in of foreign investors with technological resources and know-how. The latter is particularly relevant if productivity in the oil and gas fields is to increase. In the government's third economic plan, privatisation occupies a central place. It is recognised to offer the possibility of reversing the economic problems inherited from the past and to realise the potential for growth and development. It has been argued that at present there appears to be very little choice for the Iranian government but to privatise. Rather, the important choice concerns the modalities of implementing the privatisation policy and the industries which it will eventually cover (Shahriari-Rad, 2002).

There are a number of pronouncements and reports supporting such a line of reasoning. The Deputy Energy Minister, Reza Amrollahi, recently issued a statement to the effect that the Energy Ministry is acting more rapidly than other ministries in ceding its affiliated companies to the private sector: "We have prepared almost all subsidiary companies of the Energy Ministry for cession to the private sector and our speed is unparalleled by other state bodies" (PIN, 2004). It seems, however, that the promise for reform backfired in the parliamentary elections of February 2004, when the reformist followers of President Khatami lost their majority to conservative hardliners. As mentioned earlier, after taking office in 1997, Khatami had sponsored efforts to modernise the economy in a number of ways, including privatisation, but owing to the lack of visible progress, the opposition could not be won over.

Before moving on to the next section, there is a matter that needs to be briefly mentioned and this is the unpredictability of national politics, such as the election of Ahmadinejad which took some analysts by surprise (Pollack, 2006). More importantly for this discussion, the particular policies and policy paths that he has chosen do not necessarily align with the points made so far in this study, namely that there has been a move towards modernisation and that modernisation is a requisite of privatisation in Iran. This issue will be covered in more detail in the concluding chapter of this study, but needs to be kept in mind in the intervening pages.

The Legal and Institutional Framework[5]

While privatisation in Iran may have been initially considered in the early 1990s,[6] the first concrete indication that a privatisation policy was gaining currency within the political, economic and, possibly, religious elite in the country was the law of the Third Socio-Economic and Cultural Plan of the Islamic Republic which was passed in 2000. It incorporated two chapters devoted to privatisation: a) Chapter Two – Reorganisation of State-Owned Enterprises and b) Chapter Three – Privatisation and Management of State-Owned Enterprises. Within a couple of years of its adoption, it was supplemented by a number of by-laws regulating the implementation of some of its articles and by the articles of incorporation of the Privatisation Organisation; this was to become the main institutional instrument of the privatisation policy. Such fragmented legislation is normal where state ownership and control of the economy are pervasive. Privatisation acts usually provide only the overarching elements of the policy, leaving the details of its implementation to sub-normative regulation.

Chapter Two of the Third Socio-Economic and Cultural Plan authorises the government to take "all appropriate measures towards the divestiture, dissolving, merging and reorganising of state owned enterprises", including the "reforming and sanctioning of their article of association, and their transactional and financial by-laws" (Article 4). The stated purpose of these measures is to reorganise state-owned enterprises into specialised holding companies, managed under the auspices of resource ministries and so prepare the ground for the smooth transfer of their shares to the private sector. Chapter Two also provides for the separation of policymaking by the respective ministries from the day-to-day running of the enterprises they control. The government is also required to dissolve all overseas branches of state-owned enterprises and to co-ordinate its restructuring activities with the State Administrative and Employment Affairs Organisation.

The need for the restructuring of state-owned companies is justified with the broadly stated need to "effectively utilise their resources and potentials" (Article 4). More concretely, Chapter Two contains two related objectives: to "enhance the productivity and efficiency of those state-owned enterprises that will remain in the public sector" and to "facilitate transferring to the non-public sector the enterprises whose continued operation in the public sector is deemed unnecessary" (Article 4). The state-owned companies are guaranteed the privilege of utilising public capital, but only when they are monopolistic by their very nature, or are of no interest to the private sector to invest in. Despite its reform-oriented tone, Chapter Two (Article 5) refrains from introducing price liberalisation which, in fact, could be essential in encouraging private entrepreneurial activity into the country.

Chapter Three of the Socio-Economic Plan deals more specifically with the Iranian policy regarding privatisation. It endorses the need to "promote the role and scope of the private and co-operative sectors" in the economy and provides that "shares and stocks of the state-owned enterprises that are transferable shall be sold" (Article 9). Article 10 then sets the conditions and limitations with respect to the policy. According to Article 10, privatisation is to be considered only as a means of realising the wider objectives of the Third Plan and not as an end in itself. Furthermore, it must not endanger national security or threaten the sovereignty of the Islamic Republic, nor should it infringe upon people's rights, or create monopolies. The privatisation policy and implementation must accord with the constitution and are required to promote the widest possible public participation.

Chapter Three enumerates the holders of shares, whose transfer to private ownership it regulates. Among them are governmental agencies and ministries, as well as profit-making entities affiliated to the government (Article 11). Article 11 also mentions the Iranian National Oil Company and the shares that it holds in non-public enterprises and companies. Control and supervision over the divestiture process is vested in the High Commission of Divestiture, which was set up under the chairmanship of the Minister of Economic Affairs and Finance (Article 12). Apart from him, the Commission comprises six other members (all high-ranking government officials): the Head of the Plan and Budget Organisation, the Governor of the Central Bank, the minister concerned (i.e. representing the ministry that owns the enterprise), the Minister of

Justice, and one representative each from the Parliamentary Commission of Economic Affairs, Finance and Co-operative and the Parliamentary Commission of Plan and Budget.

The other relevant institution in the policy implementation process is the Privatisation Organisation. In fact, it replaced the Organisation for the Promotion of Ownership of Production Unit through the change in the Articles of Association of the latter (Article 15). Since it is not the owner of any state company shares, the Privatisation Organisation receives them in trust from their holders, so that it can implement the transferral policy. The organisation, though affiliated to the Ministry of Economic Affairs and Finance, is a legal entity and is financially independent. Following the launching of the Third Plan, its structure and responsibilities were further determined by the articles of incorporation adopted by a Council of Ministers Decree of 16 May 2001.

Article 16 outlined the criteria for selling the shares of state-owned companies. According to them, the shares of small companies would be sold to (individual) "entrepreneurs of high managerial calibre". Medium-sized companies were to be sold to specialised commercial groups, co-operatives and other entities. By contrast, the shares of large companies were to be sold to the general public, while preserving a controlling share for the management. In preparing privatisation of state-owned companies, the High Commission of Divestiture was required to: "determine the buyer's commitments towards employment, production programme, new investment, special activities to protect the environment, and avoidance of certain restrictive commercial activities" (Article 17).

The implementation of Articles 16 and 17 was subsequently regulated through by-laws.

In terms of privatisation revenues, Chapter Three determines the following distribution: 50% would be utilised to further finance the privatisation process; help restructure saleable companies and the re-habilitation and preparation of other state-owned enterprises earmarked for sale; as well as promoting industrial development with preference given to paying off the saleable companies' debts. Some 48% was designated to support the country's treasury. Finally, 2% was set aside for the Iranian mobilisation forces. It has to be noted that by "privatisation revenues", Article 19 does not only envisage the sale of company shares, but all kinds of contracts (rental, leasing, etc.) that generate monetary income for the owner within the given fiscal period.

Chapter Three also provides for the settlement of disputes arising from claims made by physical or legal persons against decisions embedded in the privatisation policy. Interestingly, both parties to any given privatisation deal are obliged to insert a clause to this effect in the divestiture contract (Article 20). They are thus not granted the right to choose the arbitrating body and are barred from applying to the courts; rather, they have to comply with the prescription of the Privatisation Act. Such circumscription invites questions about the reasoning behind it. There is obviously an intention to define a complete structure for privatisation relations and create what seems to be parallel institutions, such as the Arbitration Commission.

The Arbitration Commission consists of: a) five experts in economics, finance, commerce, the technical field, and law (jointly nominated by the Ministers of Economic Affairs and Finance, Justice, and the Head of the Plan and Budget Organisation and approved by the Cabinet for an appointment of six years); b) the Head of the Chamber of Co-operatives and c) the President of the Chamber of Commerce, Industry and Mines. Under Article 21, the Arbitration Commission is required to review and rule on all claims relating to divestiture. Its rules and procedures are to be formulated in a special by-law approved by the Cabinet. The minimum quorum for sessions of the Arbitration Commission is five of its members, while decisions are taken by a simple majority with on-the-record minority reports.

Once the Arbitration Commission has delivered judgement, it becomes binding ten days after the parties concerned have been notified. Within that period, any of them can raise objections to the decision and file it in writing. If an objection is made after the grace period, the parties can refer the matter to the Chief Justice of the country provided they can prove that there has been a good reason for the delay. In such an eventuality, the Chief Justice would refer the case to a special tribunal, which would then review the claim and deliver its judgement, which would become final and binding on all parties (Article 23).

Article 24 requires the government to provide insurance cover, at its own expense, for all the officials who are individually or collectively involved in the privatisation process against fines imposed on them in case of a conviction, provided that the misconduct regarding the divestiture was unintentional. The article requires the insurer to compensate the insured for whatever costs the "wrongdoer" has incurred. This is again an interesting departure from modern legal tradition where costs may be covered for lawsuits, but not in cases of a conviction. Questions arise as to why this is so in the Iranian privatisation code. One possible explanation is that the provision is designed to mitigate discontent with a process that is potentially controversial and which could result in financial loss.

Finally, Chapter Three winds up with three articles relating to liabilities of the seller and the buyer for the company's debts or for potential losses during the process of concluding the divestiture deal (Article 25), to the tax implications of privatisation deals (Article 26) and to the social insurance of employees (Article 27). Under Article 25, the divested company (i.e. its new owner) remains liable for all of its debts, which understandably should result in a discounted price, although this is not specified in the provision. Article 26 exempts all shares sold under the Privatisation Act, as well as shares transferred between executive agencies in enforcing the Act, from transaction tax. With regard to corporate income tax, this remains an obligation of the original owner up to the end of the fiscal year prior to the sale, and this effectively transfers a bigger obligation onto the buyer, especially on companies purchased in the second half of a given year.

As regards social insurance of the employees in a privatised enterprise, Article 27 provides that those who are entitled to receive a pension from the special pension funds of ministries or public agencies and whose employment with the divested company would formally terminate upon its sale to the private or co-operative sectors may, upon their reinstatement in the same company, continue to stay with the same pension fund. On the face of it this provision may seem reassuring to employees, but is in fact misleading since it does not guarantee their employment. Understandably, this can generate antagonism to privatisation within the workers' associations in any given enterprise.

In reference to the practical implementation of privatisation, the Council of Ministers decreed special articles of incorporation of the Privatisation Organisation. With headquarters in Tehran and branches throughout the country, it was charged with a number of key duties. Article 5 of the Articles of Incorporation enumerates them as follows:

> a) Formulating proper guidelines in order to expedite and facilitate public participation towards the realisation of enhanced efficiency and increased productivity of material and human resources and development of potentials of the private and cooperative sectors, and presentation of the said guidelines to the High Commission of Divestiture; b) working out plans for enhancing the capabilities of the private sector and providing incentives to specialised groups and organisations and educating entrepreneurs; c) acting as proxy in matters of divesting shares of the companies slated for privatisation on the basis of the related regulations; d) offering shares of the saleable companies according to an approved time-table; e) implementing policies sanctioned by the High Commission of Divestiture; f) rendering any other action, operation and services towards execution of the programme of divesting shares and management of governmental corporations and privatisation within the framework of the laws and regulations; g) supervising the prompt execution of the terms of contracts relating to divestment, management, rent and performance of other matters entrusted to the organisation according to the contracts; h) performing other duties related to divestment of shares and privatisation that are entrusted to the organisation within the five-year development plans.

The Privatisation Organisation consists of a General Assembly and an Executive Board with a managing director and an auditing officer, each of them having clearly defined duties and prerogatives. According to Article 7, the General Assembly consists of seven members: the Minister of Economic Affairs and Finance (President of the Assembly), the Head of the State Management and Planning Organisation, the Minister of Justice, the Minister of Industries and Mines, the Minister of Labour and Social Affairs, the Minister of Commerce and the Minister of Cooperatives. Decisions are taken by a simple majority of the members, which also constitutes the quorum for its meetings.

The meetings of the General Assembly are called by the President or at the request of the other institutions of the Privatisation Organisation. The duties of the General Assembly are set out in a ten-point list of Article 10. They are:

> a) To appoint and dismiss the chairman and the members of the Executive Board; b) to determine the salaries of the Managing Director and the Executive Board; c) to approve budgets, financial statements and the reports produced by the above two organs; d) to

> appoint the Auditing Officer and determine his/her remuneration; e) to sanction by-laws needed by the Organisation, including ones for employment, financial, administrative and business transactions; f) to adopt the general strategy of implementing the privatisation process and to approve the structure of the Organisation upon the consent of the State Management and Planning Organisation; g) to approve proxy fees and charges in connection to the sale of stocks; h) to make decisions concerning arbitration, to appoint arbitrator and to settle claims arising out of divestiture procedures; i) to propose to the Council of Ministers changes and amendments to the articles of incorporation; j) to take decisions concerning all other matters that according to the commercial law fall into the domain of competence of the general assembly, but are not determined in these Articles of Association.

The Executive Board of the Privatisation Organisation consists of a Chairman (one of the under-secretaries of the Minister of Economic Affairs and Finance), who is simultaneously the managing director, and four other members elected by the General Assembly for a renewable period of three years. In the case of resignation, retirement or demise, the vacancy is then filled by a vote in the Assembly. The Executive Board can take decisions with a minimum of three votes. The Board's duties include the following:

> a) The preparation of annual budgets, financial statements and performance reports; b) the proposal of amendments to the articles of incorporation and the organisational structure; c) the proposal of settlement of claims and reconciliation to the General Assembly; d) the proposal of necessary by-laws and fees and charges in connection to the sale of stocks; e) the preparation of strategy plans for the Organisation in implementing the privatisation policy.

Notwithstanding the powers and duties of the General Assembly and the Executive Board, the highest executive authority within the Privatisation Organisation rests with its managing director. His duties are:

> a) Implementing the policies, decisions and strategies formulated by the High Commission of Divestiture, the General Assembly and the Executive Board; b) annual budgeting in coordination with the Executive Board; c) preparation of financial statements for the Auditing Officer; d) the preparation of organisational charts,

staffing policies, annual performance reports, etc.; e) presenting the Organisation before other legal persons and judicial authorities, governmental institutions and agencies, public, private and cooperative entities both domestic and foreign, with the right to appoint attorney and to propose reconciliation and arbitration to the Executive Board; f) defining the job description for the Organisation's staff and delegating duties and responsibilities in accordance with the respective laws.

Finally, the auditing officer (also called the legal inspector) is appointed for one year according to the employment regulations in the country. Re-appointment is permissible and the articles of incorporation do not specify the number of permitted re-appointments. There is also no specification of his duties and responsibilities similar to the detailed enumeration above. Article 20 confines itself to simply declaring that his/her functions will be in accordance with the relevant laws and regulations. Thereafter, the articles of incorporation conclude with miscellaneous provisions, among which Article 22 requires the Privatisation Organisation to prioritise the divestment of the shares which are in its possession at the time of its establishment. Article 24 states that the organisation should transfer 10% of its net profit to the legal reserve account, which is required to be maintained at the level of its capital.

Analysis of the legal and institutional framework

Broadly, the legal and institutional framework of the privatisation process in Iran is outlined in the Third Socio-Economic and Cultural Development Plan of the Islamic Republic (Privatisation Organisation, 2002). On the face of it, it does not tell us much about the actual implementation of the process. As such, it raises a number of questions with respect to the background to this legal foundation, the interests and ideas that lie behind it, the modes and outcomes of the institutional interaction codified by it, as well as the practical results of actually carrying out the privatisation policy. The answers to these questions are crucial in making reliable claims about the development of the Iranian divestiture and its spread to the oil and gas sector, as well as about the opportunities available to foreign capital which would bring substantial gains to the country's economy.

In the first place, at the conceptual level, it needs to be ascertained what the likely impact of the stated objectives would be upon Iranian public enterprises and their privatisation. In other words, how does the country's elite see this relationship and what are the elements of contradiction and symbiosis between these two elements? In economics, they are not mutually exclusive, but in politics they can often turn out to be so. Another related issue is the symbiosis that the official privatisation programme stipulates between politicians and company managers. How does this work out in reality? Both groups presumably have their own agendas and the process of reconciling them could easily produce a stalemate and conflicting ideas about the practical steps to be taken.

Secondly, at a more pragmatic level, there are a number of unclear positions in the institutional framework. An important one, for example, is the exclusion of privatisation claims from the judicial system and creating a specialised Arbitration Committee subordinated to the cabinet. Would this not undermine the credibility of the whole process, or was it promulgated precisely with this objective in mind? Similarly, various lines of accountability and relations of sub- and super-ordination need to be explained and analysed. Within the institutional set-up of the privatisation process, the Cabinet stands out as a central organ, which devises the policies, modes of implementation and control, as well as the form of adjudication on claims. Why is this so? Does this not place Iranian privatisation amid political struggles, swinging it from one side to the other with every major political change?

The Cabinet is also charged with the important function of laying down the normative framework of by-laws that unpack the contents of the Third Social and Economic Plan. The articles of incorporation of the Privatisation Organisation, for example, come into effect by cabinet decree (No. H23708/T/7858 of 16 May 2001) as do the by-laws for Article 14e (Decree No. H24285/T/24094 of 13 August 2001), Article 14f (Decree No. H24653/T/23511 of 8 August 2001), Article 15 (Decree No. H23743/T/4508 of 29 April 2001), Article 16 (Decree No. H24654/T/23760 of 11 August 2001) and Article 21 (Decree No. H23833/T/45536 of 2 January 2002) of the Third Socio-Economic and Cultural Development Plan of the Islamic Republic. Such a combination of legislative, executive and judicial powers vested in one organ,[7] in this case the Cabinet, would potentially lead to closures (i.e. no-go zones and preferences) and a lack of transparency and consistency in implementing privatisation. Empirical research needs to be undertaken to establish the extent to which this is true and how this shortcoming is perceived and addressed by the Iranian political, economic and religious elite.

Thirdly, at the most basic level of defining functions, duties and powers, the researcher has to consider the broad formulation of the responsibilities of the Privatisation Organisation. The first two (see the exhaustive enumeration above) are outlined in particularly broad terms and do not remotely specify the practical task of transferring property from the public to the private sectors. What is the reason for this? Does this apply to private ownership of domestic entrepreneurs only – once they have been trained by the Privatisation Organisation? In fact, the actual selling of shares comes only fourth in the list of functions of the organisation which, anyway, seems to be overly saturated with institutional involvement, i.e. how does the General Assembly work with the High Commission for Divestiture or with the Arbitration Commission? An overarching question is whether the system is workable at all. What kind of networks within and across parallel structures need to be established, and how do the different pieces connect and control each other? Are these interrelations enabling or hindering the privatisation process?

The Oil and Gas Sector

As noted before, the Iranian economy relies heavily on oil export revenues. They form around 80% of total export earnings, 50% of the government budget and 20% of GDP (USEIA, 2005).[8] The economy's development is thus inordinately dependent on the price of oil. When it plunges, as during 1998 and early 1999, stagnation results. Conversely, when the price goes up, the government is enabled to finance economic recovery. The relatively high export revenues from oil in the past few years have allowed the setting up of an oil stabilisation fund, which provides a safeguard in the event of a fall in price. In a bid to ensure stability, the government has also attempted to diversify its activities by investing some of the oil revenues in other areas, such as the petrochemical industry.

In terms of its existing oil fields, Iran holds around 90 billion barrels of proven reserves, roughly 7% of the world's total, and claims another 30 billion barrels (so far unproved). The vast majority of the crude oil reserves are located in giant onshore fields in the south-western region (Khuzestan) near the border of Iraq and the Persian Gulf (see Table 5.1). Iran produced 3.5 million barrels in 2002 and 3.9 million barrels in 2003 (USEIA, 2005). At the end of 2003, Iran's sustainable crude oil production capacity was estimated at around 3.75 million barrels per day. It is acknowledged, both domestically and internationally, that with sufficient investment, Iran could increase its oil production capacity significantly. This can be done by improving the efficiency of the operative oil fields, or by the exploitation of new ones.

By way of comparison, Iran produced 6 million barrels in 1974, which is almost double the amount produced in 2002 and in the years following the Islamic Revolution. The Minister of Oil, Zanganeh, acknowledged the need for upgrading and modernisation, since extraction from some of the fields was carried out by methods that damage their future production capacity. The current government has devised ambitious plans to double national oil production, to more than 7 million barrels, by 2015, and is counting on foreign investment to accomplish this; possibly as much as $5 billion per year between now and then (USEIA, 2005). The production activities of NIOC are concentrated mainly on sustaining output levels from large, but ageing fields. This indicates that the oil sector is gradually becoming inefficient and is in need of modernisation through the introduction of advanced technologies and investment.

TABLE 5.1
Iranian Oil Fields and their Production Capacity

Major Iranian Oil Fields (estimated production, bbl/d)	
Onshore	**Offshore**
Agha Jari (200,000 bbl/d)	Abuzar (125,000 bbl/d)
Ahwaz-Asmari (700,000 bbl/d)	Dorood (130,000 bbl/d)
Bangestan (around 245,000 bbl/d current production, with plans to increase to 550,000 bbl/d or more)	Salman (130,000 bbl/d)
Bibi Hakimeh (130,000 bbl/d)	Sirri A&E (95,000 bbl/d)
Gachsaran (560,000 bbl/d)	Soroush/Nowruz (60,000 bbl/d)
Karanj-Parsi (200,000 bbl/d)	
Marun (520,000 bbl/d)	
Pazanan (70,000 bbl/d)	
Rag-e-Safid (180,000 bbl/d)	

Note: bbl/d stands for British barrels per day
Source: USEIA (2005).

With regard to consumption, the ratio between the exported amounts and those for the domestic market is approximately 67% to 33% (2.6 to 1.3 million barrels in 2003) (USEIA, 2005). The major overseas customers of Iranian oil are Japan, China, South Korea, Taiwan and Europe. Iranian oil is traded on the spot market by NIOC's London division. Iran's domestic consumption is increasing rapidly, which reflects the steady growth of the economy and population. The government heavily subsidises the internal prices of oil products to the value of $3 billion per year, a fact which encourages needless consumption and waste. As Iran has to import gasoline to the tune of $5 billion a year and also provides a subsidy for it, it would be sensible for the government to raise the price and introduce rationing.

In the gas sector, the scope for development is much greater. Iran has an estimated 940 trillion cubic feet (tcf) in proven natural gas reserves – the world's second largest after Russia – of which 62% has not yet been developed. Due to its enormous reserve, Iran has the potential to become a significant natural gas exporter; in 2002, for example, Iran produced about 2.7 tcf of natural gas. In March 2004, Iran signed a $1.2 billion contract with a consortium consisting of two foreign and two domestic companies to gather natural gas from the Nowruz, Soroush, Hendijan and Behregansar fields (ibid.). Russia's Gazprom has also expressed interest in developing Iranian gas resources. Currently, natural gas accounts for nearly half of Iran's total energy consumption, and the government plans billions of dollars worth of further investment in the coming years.

As concerns the privatisation issue, the Iranian constitution (adopted following the Islamic Revolution) currently prohibits the granting of petroleum rights on a concessionary basis or direct equity stake. In 1987, the Petroleum Law was passed permitting the establishment of contracts between the Ministry of Oil, state companies and entities, whether personal or legal, local or foreign. So far, the most often used system is the so-called buy-back agreement whereby the NIOC concludes contracts with companies willing to invest in the development of new oil fields or in the improvement of existing ones. Once in place, the buy-back mechanism confers the obligation on the companies (domestic or foreign) to fund all necessary investment for the project and the right to receive remuneration from NIOC. On the expiry of the agreement the operation of the fields reverts to the NIOC which now retains the whole revenue from sales.

Throughout the years of its implementation, the buy-back arrangement has been assessed as having both drawbacks and advantages for each of the contracting parties (personal communication).[9] The NIOC has no guarantee against a possible drop in oil prices and has to pay the remuneration to the contractor anyway. So it bears the whole risk in the initial years of the project, and might possibly make payments that outstrip its revenues. The advantage is that it remains the sole proprietor of the field and, as already mentioned, retains all of the revenue after the expiry date of the contract. As for the investors, they have no guarantee that they will not be barred from developing the oil fields that they discover after the initial investment. The current political system of Iran bears this risk alone. Against this, however, investors bear no risk of falling oil prices as they still receive the agreed payment for their work. However, if oil prices fall, the contractors will receive their payback over a longer period, in effect receiving less financial reward. So the contractors also bear some risk of a fall in oil prices.

The current political risk in Iran stems largely from the antagonism between the conservative and reform-oriented forces. The former are represented by the Council of Guardians (a 12-strong religious body formed after the Islamic Revolution), which has veto power over laws passed by the National Assembly and also supervises the elections. The latter formed the base of President Khatami's overwhelming electoral victories in 1997 and 2001 and were represented by his government. The conservative forces led in the implementation of the nationalisation policies during the post-revolutionary period. As such, they vehemently opposed any efforts to liberalise the economy or to offer state-owned enterprises to foreign investors and buyers. This was exactly what the reformists wanted to do, arguing that this was the only way the Iranian economy could recover in a sustainable way (Katouzian, 1989).

The conflict between the two opposing sides manifested itself quite visibly on at least two occasions. In May 2001, the National Assembly (then dominated by reform-oriented MPs) passed a bill designed to attract and protect foreign investment. The following month, the Council of Guardians rejected it as being too liberal. Heavily amended, the bill was passed again by the National Assembly in November 2001, but a month later the Council of Guardians rejected it for a second time. Following a number of consultations, the now greatly weakened bill was finally adopted in May 2002 and approved by the Council of Guardians as the

Law on the Attraction and Protection of Foreign Investment. It aimed to encourage foreign investment by streamlining procedures and guaranteeing profit repatriation. In fact, the law, which was sent to the government for implementation in January 2003, was the first act regarding foreign investment passed by Iran's legislature since the Islamic Revolution.

Similarly, the conservatives have proved themselves potent in frustrating the economic policy of Khatami. For example, they introduced retroactive legislation demonstrating disdain for private property rights. After winning the February 2004 elections for the National Assembly, the conservative majority passed a bill which would allow them a veto over foreign investment issues. The Petroleum Bill was adopted in September 2004, but was backdated to March 2004. Under its provisions, foreign holdings of Iranian enterprises had to be approved by the Assembly. Its backers did not conceal their willingness to employ this piece of legislation to terminate two deals with Turkish consortia on the grounds that they had previously conducted business with Israel and were thus considered to present security risks for Iran. The first was the Turkish–Austrian group Tepe-Akfen-Vie, which had the concession to build Tehran's new Imam Khomeini airport. The second was Turkcell, which had earlier been awarded a $3 billion contract to build a second GSM network in Iran (BBC, 2004a).

Following Khatami's victory in the presidential elections of 1997, the NIOC and the Ministry of Oil concluded a number of buy-back contracts with large foreign investors in the oil sector. Despite the fact that some of them had been tainted by corruption scandals involving foreign oil companies and high-ranking Iranian officials, they did bring considerable investment to the country. This policy reflects the reformists' commitment to the economic development of Iran and is part of the political struggle in the country. Another of the restrictive acts of the current conservative majority in the National Assembly was the rejection of a proposal to give oil exploration companies the right to exploit their discoveries.[10] Such proposals formed a central plank of Khatami's liberal economic plan, introducing privatisation and attracting more foreign investment. Currently, there are a number of buy-back projects being implemented in Iran, several of which are presented below (USEIA, 2005).

A final agreement was signed in February 2004 on the $2 billion Azadegan oil field development project, with the field being developed by Iranians (although Inpex is said to hold a 10% interest).[11] Azadegan was discovered in 1999, representing Iran's largest oil discovery in 30 years, and is located onshore in the south-western province of Khuzestan, a few miles east of the border with Iraq. Reportedly, Azadegan contains proven crude oil reserves of 26 billion barrels (USEIA, 2005), but the field is also considered to be geologically complex, making oil extraction more challenging and expensive. In January 2001, the Majlis approved the development of Azadegan through a buy-back scheme with a foreign investor. As the biggest importer of Iranian oil, Japan agreed to loan Iran $3 billion in exchange for receiving the development rights for Azadegan. According to estimates, the output from Azadegan alone could, at its peak, account for as much as 6% of Japan's oil imports.

Since 1995, NIOC has made several other sizable oil discoveries, including the 3–5 billion barrels in the Darkhovin onshore oil field, located near Abadan. In late June 2001, Italy's ENI signed a $1 billion, five-and-a-half-year buy-back contract to develop Darkhovin. The contract contained a clause reducing the risk for the investor. The added incentive included a limited risk/reward element, whereby the payment of NIOC remuneration was linked to the production capacity of the oil field. The provision aimed to neutralise the disadvantages inherent in the buy-back model and as such represents a significant improvement of the Iranian oil development policy. Ultimately, the production capacity at Darkhovin oil field is expected to reach 160,000 barrels per day, which ranks it relatively high among other production sites.

The first major project under the buy-back investment arrangement became operational in October 1998. The Sirri A oil field was developed and operated by Total and Malaysia's Petronas and began production at 7,000 barrels per day (USEIA, 2005). The neighbouring Sirri E field began production shortly thereafter (in February 1999). In March 1999, France's Elf Aquitaine and a consortium of Italy's ENI and Agip were awarded a $1 billion contract for the recovery of the Doroud oil and natural gas field. The project was to boost its production capacity by approximately 50% by 2004. In April 1999, Iran awarded Elf (46.75% share), along with Canada's Bow Valley Energy (15% share), a buy-back contract to develop the offshore Balal field. ENI was also involved, with a 38.25% stake. The field, containing some 80 billion barrels of reserves, started producing at a rate of 20,000 barrels per day in early 2003, and reached 40,000 barrels per day in February 2004 (ibid.).

In November 2000, Norway's Statoil signed a series of agreements with NIOC to explore for oil in the Strait of Hormuz area. The agreement envisaged co-operation on the development of a processing plant for natural gas from four onshore fields and the possible development of the Salman offshore field at a cost of $850 million (ibid.). On 18 March 2005, a much sought-after contract to develop the Bangestan reservoirs in the three giant fields of Ahvaz, Mansuri and Abe Teymour was signed. Bangestan projects were awarded to Petro Iran Development Co. (a subsidy of NIOC), after having been delayed several times since 2001. Bangestan contains an estimated 6 billion barrels of oil reserves, but the field is one of the oldest in the country, requiring investment and technological applications to compensate for natural decline. Shell, Total and BP had been bidding for the project, but stated they were frustrated with the slow pace of negotiations and the numerous changes to its terms (ibid.).

In May 2002, Iran's Ministry of Oil signed another buy-back contract (for $585 million) with the NIOC subsidiary PetroIran to develop the Foroozan and Esfandiar offshore oil fields (ibid.). PetroIran was expected to increase their production at the fields to more than 120% by the beginning of 2006. In January 2004, the Cheshmeh-Khosh field, which had previously been awarded to Spain's Cepsa for $300 million, was re-awarded to the state-owned Central Iranian Oil Fields Company (CIOFC). In December 2003, Cepsa and OMV[12] withdrew from lengthy negotiations after a reported failure to agree on the development costs and the terms of the buy-back project. However, both companies had stated they could still be involved at the site in some way under conditions acceptable to them.

Similarly, Iran applied the buy-back model to the development of natural gas resources. In September 1997, Total signed a $2 billion buy-back deal, together with Russia's Gazprom and Malaysia's Petronas, to develop part of South Pars (Iran's largest gas field, part of what is likely the largest gas field in the world). Total has a 40% share of the project, with the other two companies each having 30% (USEIA, 2005). Further on in the phases of the development project, ENI and Petropars will also be involved in the construction of onshore treatment facilities. Hyundai has also signed a $1 billion contract with NIOC to build four natural gas-powered trains. Shell and Statoil have also expressed interest in participating in the development of the South Pars gas field. Also active has been a consortium of national and international companies led by Norway's Aker Kvaerner.

In recent times, the Iranian government appears to have had some doubts about the viability and effectiveness of the buy-back projects. Among such are: a) recurring charges of corruption of high-ranking officials in the Oil Ministry and NIOC and its subsidiaries; b) insufficient benefits to Iran and its economy and c) worries that buy-backs are attracting too little investment. There are signs that the reformist political elite is considering substantial changes in the system (personal communication).[13] In late May 2002, Canada's Sheer Energy became the first foreign company since the Darkhovin deal to reach an agreement under the terms achieved by ENI one year earlier. However, even this re-introduction of the limited risk/reward model has not attracted the amount of foreign investment that the Iranian economy needs.

Analysis of the oil and gas sector

It seems clear that the most important issue to take account of in an examination of what the determinants are of Iran's privatisation policy in the oil and gas industry, is the sheer size and significance of the oil and gas sector. The previous section, combined with the historical chapters (2, 3 and 4) illustrated this matter to a certain degree. Oil has been the lynchpin of Iran's economic life and has had an overwhelmingly important role to play in determining the nature of the Iranian economy, its political system and even its international relations. Although policy may seem to determine the nature of the privatisation of the oil and gas industry, in many ways it is oil itself that determines, and will continue to determine, the nature of the policy – at both the foreign and domestic levels. Chapters 2, 3 and 4 described the various ways in which oil has played a fundamental role in Iranian domestic and foreign policies and policymaking.

Beyond the size and prominence of the Iranian oil and gas industry to Iran itself, we must also consider its importance to the world in general (Klare, 2005). It is true that only Saudi Arabia possesses more oil than Iran – 260 billion barrels to be specific. Even Iraq, against which some say the United States fought a war mainly for the strategically important oil reserves, lies behind Iran with its 115 billion barrels of oil in reserve. In total, it is estimated that Iran has a full one-tenth of the world's total supply of oil (USEIA, 2005). These figures cannot be underestimated, although it should be borne in mind that estimates of total reserves can vary over time, depending on the accuracy of measurement, discovery of new fields among other factors.

One should not forget that Iran has 940 tcf of natural gas reserves (PIIC, 2000), representing some 16% of the entire world's reserves of natural gas. The only other state with larger reserves is Russia with 1,680 tcf. Translating the gas supply in terms of oil may help to demonstrate just how massive a resource gas is. In this connection Klare notes:

> As it takes approximately 6,000 cubic feet of gas to equal the energy content of 1 barrel of oil, Iran's gas reserves represent the equivalent of about 155 billion barrels of oil. This, in turn, means that its combined hydrocarbon reserves are the equivalent of some 280 billion barrels of oil, just slightly behind Saudi Arabia's combined supply (Klare, 2005:1).

In terms of combined hydrocarbon resources, Iran is one of the most important, if not the most important, countries in the world.

At the same time, one must be careful not to consider only the actual amount of oil and gas reserves in our calculations, but also the relative future production capacity of Iran versus other leading oil and gas producing states in the world. Analysts note that while Saudi Arabia has higher reserves than Iran, the former is currently producing at near maximum rate (see, for example, Klare, 2005). This is estimated at around, or over, 10 million bpd. Within the next two decades, it is unlikely that Saudi Arabia will be able to exceed that rate by much.[14] Meanwhile, the oil consumption of the world's leading oil-consuming countries – the US, China, and India – is estimated to rise by 50% in that time. In other words, world demand for oil in the next two decades will not be met by Saudi Arabia's supply. Within the context of supply and demand in the world system, Iran has a significant potential to increase its yearly production of oil. Currently, Iran is producing around 4 million bpd (PIIC, 2000), but is capable of increasing output to 7 million bpd. This fact alone puts Iran in a league of its own in terms of its potential for future oil production.

Iran's production of natural gas is nowhere near maximum. At present it produces approximately 2.7 tcf per year and, given the huge size of its reserves, it is clear that Iran has a huge potential to increase its exploitation (Koussari, 1999). Furthermore, it is one of the few countries which enjoys such a potential, again putting it in a league of its own.

These statistics point to the degree to which oil and gas influence policy rather than the other way round. There is significant pressure from all parts of the world for Iran to release its supplies of oil and gas. On this basis, oil and gas provide Iran with an extremely powerful bargaining chip in determining its foreign policy. The sheer value of its reserves to the rest of the world, in particular to the world's most powerful countries (which also happen to be the countries which are most dependent on oil) gives Iran considerable room for manoeuvre in negotiations with foreign powers. This manifests itself clearly in the negotiations over the nuclear issue. On the other hand, the value of Iran's hydrocarbon wealth is a double-edged sword, for it puts the country under constant threat from states which fear for the safety of the world's oil and gas supply.

All told, at least since the revolution, Iran has continued to treat oil as a strategic resource and as such, it is bound to shape the nature of the privatisation of the oil industry, making it different from any other type of privatisation. Iran is not alone in viewing oil and gas as strategic resources and protecting them is therefore of great importance, deserving special privatisation strategies. As noted in Chapter 1, other states like Egypt and China share this view. It is in this context that the discussion in the previous section about buy-back arrangements regarding the privatisation of oil and gas must be considered.

The policymakers' views

There is an urgent need to reform Iran's economy, particularly since the collapse of the Iranian stock market. This was essentially caused by Mahmoud Ahmadinejad's call to "wipe Israel off the map" coupled with his announcement, in November 2005, to the effect that the general public would receive shares in publicly held industries. This was taken to be an initial step towards the seizure and redistribution of the many private industries operating in Iran. As a result, there was a sudden and massive flight of capital from the country, perhaps amounting to around $200 billion.

On his election to office in June 2005, Ahmadinejad stated that he would stamp out corruption. However, around 40% of Iran's GDP is accounted for by the Bonyads, which, it may be remembered, were essentially set up as charitable foundations, but were in actual fact massive corruption machines, bankrolling the senior leadership. Such corruption resulted in increased inflation and unemployment, reduced liquidity and investment, diminished non-oil exports and further widened the gap between rich and poor.

Consequently, Iran called for some $20 billion investment each and every year, in addition to $70 billion to recapitalise the country's decrepit oil industry. One major problem was that there were only three capital markets in the world capable of such high investment levels – the United States, Europe and Japan (Amini, 2004; Nili *et al.*, 2004). Furthermore, these are unlikely to be replaced in the foreseeable future. Also to be considered is the far superior technology available to Western companies (Pollack, 2006). During his presidency, Khatami realised the need for market reform and, in the absence of local capital, advocated ways of attracting foreign investment. He applied a short sharp shock to the market through the privatisation of some state firms, and embarked on a policy of liberalisation, allowing more freedom of expression to the people and to the popular press, as well as for some changes in labour laws.

President Khatami's government approved a massive programme of privatisation of state-owned companies (Reuters, 1999). Parliament also passed plans to modernise Iran's oil industry and almost double its production of oil over the next decade (USEIA, 1998). Such an increase in production required massive investments (some $120 billion), and so Iran looked to local private investors and foreign sources for capital. (Iran has not yet discovered oil on the Iranian side of the Caspian Sea, but the potential is there.)

The final and immediate determinant of Iranian privatisation policy regarding the oil and gas industry is the personal, subjective views of those policymakers and policy-enactors who are concerned directly with, or are otherwise relevant to, privatisation. To garner these views, I conducted extensive primary research and assessed them. Two methods were employed in this endeavour; the first was by means of a questionnaire to be filled in by the personnel concerned, and the second by holding face-to-face interviews with a number of them, mostly key figures. In many ways, the human factor is one of the most important to take account of because its effect is immediate in determining policy. In a way, the answers of the interviewees may be thought of as providing an answer to the question of this study as a whole, as it may show the human element to be the most important, perhaps even the only, determinant of the oil and gas privatisation policy now and in the future.

Yet there are multiple problems and issues which arise in examinations such as this, making such a simple answer somewhat untenable and the views of these individuals as representing one factor among many. These problems can be seen as involving, at their very base, epistemological and methodological issues. Some of the epistemological issues have been discussed in the section above. Specifically, it was noted that there has been debate concerning the degree to which historical and other structural factors truly affect the decisionmaker's final decision. This issue comes up again in this section, although in another form. This time, however, the issue is that the determination of policy is not necessarily made in terms of what people think they think. Individuals may report that they believe they will base their decision on a set of particular views, historical experiences etc., and then may completely ignore such views and make a decision using other methods. This means that for the analyst, knowing these views, historical experiences and so on, may not necessarily lead to a prediction of the decisionmaker's action.

A related issue is that such misreporting is particularly prevalent in the realm of politics. This is because decisions are made not only in terms of the policymakers' own views, but also in terms of their constituency. Such a constituency may involve the people who elected them to office, those within a bureaucracy they work in, or those to whom they must respond, such as religious personalities. Thus, the personal views of someone involved in politics may not necessarily lead to the actual policy to be implemented. At the same time, the political individual may misreport his or her views in order to please his or her constituency. Thus, one must take the results of such personal inquiries with a grain of salt.

These different epistemological issues force us to rely on other factors in order to answer the central question of this study: what are the determinants of oil and gas privatisation policy in Iran? It is for this reason (the idea that the results of simply asking people what they think will not always answer your question), that the study has taken a multi-pronged approach in examining the determinants of privatisation policy. It has included theory, history and contemporary occurrences in addidtion to the subjective views of policymakers.

Related to these epistemological issues are methodological issues, which concern the nature of the analysis being made (questionnaires versus interviews). There are also issues concerning the means by which these methodological tools and their findings are analysed. The following section addresses these issues, before turning to an overview of the actual results and findings from the primary research.

Constructing the interview questions

In any interview-based research there will always be bias. This is due to the inherently biased nature of human interaction and, in a deeper sense, to the inherent bias in human thought. In the study of foreign policy, the field of international relations specifically addresses such issues. An example is the bias brought on by bureaucratic placement. The old saying, "where you sit determines where you stand", sums up the problem well – individuals are recruited and then initiated not solely into a profession, but also into a professional outlook on how problems should be solved; and this determines their response to questions.

However, the bias can be controlled. One way to control bias is to be very specific and frank about who is answering what question and how they answer it compared to others in a different situation. Of course, there are issues of confidentiality in providing the names of interviewees. However, it is generally acceptable to name the organisation for which an individual works and, sometimes, their position. In this study, I have been able to provide such information. Thus, in the following sections, all responses will be tagged to individuals based on the organisation to which they belong. In order to safeguard the confidentiality of the replies from each of the individuals interviewed, or who replied to the questionnaire, the full and complete transcripts from each interview and questionnaire are not included in this book. Another issue is that interviews often provide "spotty" information (Robson, 1997). The limitation on time spent with an interviewee in addition to the knowledge of any one particular interviewee limits the number of responses that can be obtained. It is therefore necessary to control this.

In this study, the entire inventory of questions was submitted to each interviewee. This was done in the belief that an interviewee's inability to answer a question can tell us something about his view, or his bias in favour of the organisation he works for. For instance, it may be found that those in a particular bureaucracy are unable to answer questions about the public's perceptions of privatisation as a rule, whereas in another bureaucracy they may have more insight. This could then tell us something about the sensitivity of one bureaucracy to the public's perceptions relative to another. It also limits the bias of the interviewees by leading them to the same thought pattern as other interviewees. For instance, it may be the case that an interviewee will respond to a question concerning the public's perceptions of privatisation if at the outset oil and gas are mentioned as the topic of the research, otherwise they may simply talk about privatisation in general. Finally, I believe that by asking questions, sometimes I just got lucky.

The list of questions for the interviewees was divided into sets of questions concerning the following subjects: background and privatisation policy, the legal framework, implementation, outcome and follow-up, the public's perceptions of privatisation, political influence, and the oil and gas sector specifically.

List of interview questions

Background to the privatisation policy

- What have been the main debates leading up to the formulation of the Iranian privatisation policy? At least three categories of debates must be analysed: academic, political, and economic.
- Were there differences of opinion between the team in charge of the actual drafting of the law and the Cabinet? If yes, on what issues?
- To what extent has expert opinion played a role? Or was it largely marginalised in favour of political considerations? What was the profile of the expert group in charge of drafting the privatisation law?
- Which actors (politicians, religious elite, interest groups, formal and informal associations, etc.) were most involved and provided most weight to the drafting of the initial privatisation legislation and its subsequent by-laws and amendments?
- Does the privatisation framework favour a particular social, economic, political group? Does it make clear a distinction between domestic and foreign capital? If so, why?
- Why did so much time elapse between 1995 (when the privatisation programme of Iran was officially inaugurated) and 2000 (when the Third Plan actually codified procedures and institutions)? What were the reasons for this delay?

- In those five years who were the most active actors in pushing for the actual privatisation codification and the go-ahead for the Iranian privatisation programme by the Third Plan?

Legal framework

- Did the privatisation legislation regulate the most important issues? If not, what were the areas that were insufficiently regulated or omitted altogether? How did this change at later stages?
- Consistency of legislation: are the by-laws decreed subsequently by the Cabinet in accordance with the general framework set out in the Third Plan? Who were the initiators and the most influential actors in devising the by-laws?
- What are the main areas allowing for discretion of key institutions and officials in charge of carrying out the privatisation programme? How has this discretion been exercised and to whose benefit?
- Within the legislative and institutional set-up who has come to be the central and most influential institutional actor? Why?

Implementation

- What is the level of political control and interference in the privatisation process? Are there concrete examples of how this was done in practice?
- Methods of privatisation: who is in charge of selecting one from among various privatisation methods? Which ones are the most frequent types used? What are the most common types of preferences and closures?
- Who is responsible for evaluating assets of companies identified for privatisation? Has this practice been abused?
- Are there detailed criteria that have to be met for the privatisation deal to be approved?
- What was the most common form of breaching the legislation on privatisation? How are such cases dealt with?
- What is the practical impact of the relationship between enhancing economic efficiency and privatisation, which is clearly stated in the laws and regulations?

Outcome and follow-up

- Who are currently the main winners and losers in the privatisation process?

Public perceptions of privatisation

- To what extent was privatisation something of importance to local people?
- Has this changed during the past ten years?

Political influence

- Has the direction of the privatisation programme changed with changes in government?
- Is there any political purpose served by the oil and gas sectors, i.e. is it a politically charged industry in the sense that the way it is being regulated creates dis/advantages to different political parties?

The oil and gas sector

- How much has already been privatised in the Iranian oil and gas sector?
- What trend do statistics reveal, i.e. in which direction is Iran moving – a global strategy for the whole sector or compartmentalising it and deciding different strategies for each separate area?
- Has the legal and institutional framework been devised to account for substantial privatisation and modernisation in the oil and gas industry?

As is apparent from the above questions, the interviewees were each given a raft of issues to comment on. Of course, all interviewees did not comment on each issue at length. However, the response rate was high enough to draw a number of conclusions. The following section will detail the findings of these interviews.

Findings

A total of 50 people were contacted and discussions were held with 32 of them. The interviewees included representative members of industry, government and academia. (The full list of respondents is provided in Appendix 1.) A total of 100 questionnaires were sent out to representative members of the government and academia in Iran. Only seven completed forms were returned. In some cases the questionnaires were filled in and the individual was interviewed. The low figure of unanswered questionnaires showed that face-to-face interviews worked better in this case. Appendix 2 gives a list of those who returned the questionnaires.

Based on these interviews and the questionnaires which were answered, it is possible to draw some generalisations and conclusions. Most of this study has taken a structural approach to the analysis of factors influencing the privatisation policy in Iran. By generalising the results of these interviews, it is possible to add to this structural approach (Silverman, 2000). This is done through a process of detailed comparison of interviewees.

The findings can be roughly split into two categories. First, there are the perceptions of why, how and when privatisation will take place and what type. Second, there are the perceptions of what groups influence the privatisation policy and how they go about doing so.

The type of privatisation: the whys and wherefores

Some interviewees (see Appendix 1) suggested that there were structural factors involved in the decision to privatise in Iran. In particular, the relative price of oil is thought to play a major role in the possibility, or at least the pace, of the privatisation of the oil and gas industry. One interviewee remarked:

> There has been a direct correlation between lower oil prices and faster pace of privatisation. Every time the price of oil has dropped in the international market, the pace and the impetus for privatisation has picked up. When in 1989/1990 the talk of privatisation started, right after the Iran–Iraq war, Iran hit the bottom of the selling price for its exports of oil. When this phenomenon is present, the government thinks seriously about reducing its cost of running its vast operations and therefore thinks of divesting itself of the companies it operates (Interview with Dr Parviz Aghili, Managing Director, Karafarin Bank, Tehran, 29 December 2005).

Whatever the structural issues involved, it is generally assumed amongst most members of the elite that the process of privatisation would make the Iranian economy more efficient and profitable. Furthermore, it is assumed that this process may also lead to better prospects for the treasury, in that it would increase its revenue from a burgeoning economy and have less need to pay out. The following quotes were taken from different interviews:

> The effect of privatisation [on the economy] will improve the profitability of the entity and hence will benefit the workers and the shareholders alike. Furthermore, this will reduce the size and the burdens of the government enabling it to regulate and supervise the activities, to everyone's benefit (Interview with Mr Houshang Naderian, Chairman & Managing Director, Audit Organization, Tehran, December 2005).

Other benefits noted were that privatisation would

(a) abolish all governmental restrictive policies and procedures enabling the management to act quickly for the benefit of the organisation;

(b) reduce the size of the government and hence its overall expenses;

(c) distribute the assets and holdings amongst ordinary people;

(d) increase profitability (Interview with Mr Hossein Mamdooh, Member of Board of Directors, Iran Helicopter Services Company (IHC) and Head of Quality Assurance, Tehran, December 2005).

Although there is a clear logic behind the drive to privatise, most interviewees also noted that there were significant delays in implementing privatisation. While some made a point about structural effects, such as the price of oil, others simply noted that there were inefficiencies in the system:

> The loss of time and the delays were the direct result of the lack of knowledge and clear-cut policies of the officials at the start. The firms that were privatised initially ended up sacking its workers and ended in the closure of those production units and, hence, created the delays in establishing clearer ideas and methodologies for privatisation (Interview with Mr Hossein Mamdooh, Member of Board of Directors, Iran Helicopter Services Company (IHC), Tehran, December 2005).

These kinds of inefficiencies are to be expected given the size of the undertaking. However, it was not clear what the relationship was between these inefficiencies and other structural effects, nor was it clear which issues the elite considered to be more causal. One factor that came up was that there were too many controls and interest groups involved. One interviewee noted:

> There are too many controls and interferences. This is the reason for the slow pace with which this process has moved forward since its inception. A case in point is when the oil ministry decided to sell off its drilling activities and privatise this division. Economically it was a viable proposition but since it was not accepted politically, this plan was set aside (Interview with Dr H. Ghanimi Fard, Executive Director, International Affairs, NIOC Petroleum Ministry HQ, Tehran, 22 December 2005).

Another factor is that changes in government have led to changes in privatisation policy, as expected. One source remarked:

> Unfortunately, every time a new government takes over, the privatisation programme goes through some modifications. It is as if they wish to re-invent the wheel, by putting their own signature on the agenda. By making the alterations and refitting the programme to suit their personality and to have their own signature on the scheme, much valuable time has been lost in the past (Interview with Mr Iraj Nadimi, MP, Deputy Head of Majlis Economic Committee, and university lecturer, Tehran, 24 December 2005).

Although many interviewees seemed to agree that privatisation was a necessity and that at some point there would be a certain degree of privatisation of the oil and gas industry, there was not always agreement on which parts of the industry would be privatised. However, one of the recurrent views was that the downstream activities of the oil and gas industry should be privatised before those upstream. For instance, one interviewee noted:

> It seems to me that the downstream activities [marketing, distribution, transportation, refining] are more prone to being privatised in the short term (Interview with Dr Narsi Ghorban, Managing Director, NarKangan (NGTLC) Gas to Liquid International Co., Tehran, 18 December 2005).

One factor in particular is the capital-intensive nature of an upstream industry. One interviewee commented:

> I do not feel we are going to see a move in the upstream side of the oil industry being privatised as yet. There are a number of sections in the downstream side of the oil and gas industry that have already been privatised, and shall continue this trend. However, there are limits to privatising capital-intensive units i.e.: the refineries which are being heavily subsidised. The distribution, tanker company, and some others have already been sold off to the oil company pension fund organisation, before they go onto the Tehran Exchange, for a full-blown privatisation and IPO (Dr H. Ghanimi Fard, Executive Director, International Affairs, NIOC Petroleum Ministry HQ, Tehran, 22 December 2005).

Another interviewee said that politics might play a role in the privatisation of upstream versus downstream sectors:

> It does not appear to be a politically intensive industry, but it is. There are politicians who wish to stay close to the industry and this naturally breeds politicians trying to place their own close associates to run sections of this industry. I do not think that the politicians are going to have much effect on the upstream side of the industry. However, they probably will exert pressure in the downstream side of oil and gas, through their contacts and connections that they try to promote into positions of power in the sector (Interview with Dr Mehdi Assali, President, Institute for International Energy Studies (IIES), Tehran, 19 December 2005).

For and against

The question of politics leads naturally to the issue of what groups within Iran are for or against privatisation. In fact, the general consensus was that oil and gas privatisation in Iran is fundamentally a political issue. As one interviewee remarked,

> petroleum and all its by-products are politically influenced and politically intensive commodities. Naturally, the industry is also motivated by the political agenda. There are certain people who could profit by keeping the oil sector this way (Interview with Mr Hassan Lajevardi, Lecturer and, Deputy Director Economic Research, Centre for Energy Studies-MATN, Tehran, 20 December 2005).

Put more boldly, which are the interest groups involved? The interviews and the questionnaire made it possible to get some idea of who the elite perceive to be interest groups. Different interest groups have different views on the issue and each group attempts to bring about its choice of policy. Yet, most interviewees seemed to believe that the privatisation plan did not favour one group or another. However, they also tended to believe that it was the lack of transparency and the complexities of the process which has "left a bad taste" in some people's mouths. For instance, one academic said:

> From the looks of it, [privatisation] does not appear to be favouring a particular segment of the public. At the beginning the implementation of some of the production units took place by direct negotiations. This left a bad taste for many who were watching, whilst the de-nationalisation of these assets ended up in favour of the political elite's cronies (Interview with Dr Ebrahim Razaghi, Professor of Economics, Tehran University, Tehran, 26 December 2005).

Nevertheless, although the privatisation plan may not specifically favour one particular group over another, there are clearly winners and losers in the privatisation plans in Iran, and this has created interest groups on both sides who bring pressure to bear on policymakers. They do not necessarily make the decisions on privatisation directly, but nonetheless, they seem to have some effect on the decisionmaking process, at least in the eyes of the elites.

Broadly speaking, some sources say that well-connected people and those involved in the political system are the beneficiaries of privatisation.[15] It is no surprise that the political elite is generally thought to be in one mind to back it. As one interviewee put it:

> There is harmony and consensus amongst the political elite. The differences are usually pronounced in some using privatisation as a tool to balance their yearly budgets, whilst others think of it as the engine for economic progress and development. These two differing opinions are probably still in place (Interview with Mr Houshang Naderian, Chairman & Managing Director, Audit Organization, Tehran, December 2005).

While the political elite is perceived as being in favour of privatisation, the higher echelons in the civil service were disgruntled by the whole idea. Of course, not all the civil servants shared this attitude as evidenced by the words of a Deputy Minister, no less, who said:

> The biggest problem of the consecutive governments has been that they have become 'competitors of the people', rather than passing on the management of the companies to the people. The ministries who have been asked to divest themselves of the companies under their wing are reluctant as they see this as a reduction of their 'reach, or their powers' (Interview with Engineer Mohamed Hossein Nejad Hosseinian, Deputy Minister of Petroleum, International Affairs, Tehran, 18 December 2005).

Another prominant figure at the interview commented:

> [The] winners are the employees as a whole and the selected capable management of the newly privatised institutions. The losers are the narrow-minded old government managers who have taken over the privatised institution and are too afraid of taking risks or bringing about new ideas which could improve the competitiveness and profitability (Interview with Mr Hossein Mamdooh, Member of Board of Directors, Iran Helicopter Services Company (IHC), Tehran, December 2005).

Like the civil servants, those on the left of the political spectrum have tended to oppose privatisation. For instance, one interviewee noted:

> Yes, there were those leftist-oriented cabinet ministers who were against the privatisation and deep down inside did not wish to see the plan for privatisation succeed. Whilst they gave lip service by saying yes to the concept, in reality, they were not believers in the programme (Interview with Dr Parviz Aghili, Managing Director, Karafarin Bank, Tehran, 29 December 2005).

As might be expected, such views were expressed by those who were already involved in privatised businesses (i.e. by those on the right). The unions felt uneasy about privatisation because of possible redundancies. This is what one interviewee said:

> There were no direct objections to the privatisation plans, except, we have seen the unions and middle management were putting stumbling blocks on the way of implementing this programme. They were afraid of losing their jobs. These same workers were those who helped to bring about the revolution and therefore, in the current labour laws, out of 203 Articles, 102 Articles are in place to protect the workers against the owners! The labour laws have to change if

> privatisation is to succeed (Interview with Dr Ali Hamadani, Legal Advisor, Society for Industrial Management, Tehran, 17 December 2005).

The interviewees also spoke about the attitude of "the masses" toward privatisation. It is generally assumed amongst members of the elite that privatisation is something which mainly affects and concerns them. The masses are thought to take little interest in it and simply do not care very much about it one way or another.

> The masses, i.e. blue-collar workers, farmers, and generally the common people, are not too bothered by this concept. The academics, management people, upper middle classes with knowledge of economic issues, this concept is considered to be very important (Interview with Mr Houshang Naderian, Chairman & Managing Director, Audit Organization, Tehran, December 2005).

One reason for this, commented another, is that

> [the] public has not had a choice or a say in the matter. They do not know how to differentiate between the private and public sector entities. [However] . . . the public is gradually becoming more educated about differentiating between the private and the public sector (Interview with Dr Parviz Aghili, Managing Director, Karafarin Bank, Tehran, 29 December 2005).

In some cases, the interviewees pointed out that the masses were opposed to privatisation. For instance, one interviewee, the Deputy Minister of Petroleum, remarked:

> The common people and the masses do not think favourably of this concept. They think here is another way for the 'rich to get richer' ploy. They think of this issue as yet another way of auctioning away what belongs to future generations. The reason is that they have yet to see a successful outcome from the entities that were privatised (Interview with Engineer Mohamed Hossein Nejad Hosseinian, Deputy Minister of Petroleum, International Affairs, Tehran, 18 December 2005).

In particular, it was thought that in some cases

> the local people see this process going against the distributive policies of the government. To this end, they are not in favour of this agenda (Interview with Dr Abbas Maleki, Chairman, International Institute for Caspian Studies, Tehran, 24 December 2005).

Although the interviewees noted that there were several different constituencies which were either for or against privatisation, they all tended to know who were specifically responsible for designing the privatisation policy. For example, an interviewee put it concisely as follows:

> The Privatisation Organisation is charged with this task. However, the Central Bank of Iran, Economic Committee, Tehran Stock Exchange and MPO (Management and Planning Organisation) can all have inputs into the discretionary issues raised on privatisation of the companies belonging to the government sector. They could also present a bill through the Cabinet of Ministers, and once ratified by the Majlis and the Expediency Council, the Privatisation Organisation will implement the execution (Interview with Mr Ali Khorram (former Ambassador to China), Deputy, High Council of Free Trade – Industrial Zones Secretariat, Tehran, 17 December 2005).

A related point is the fact that some of the respondents also addressed the question of whether a change in government would affect the course of privatisation in Iran. As this study has tended to argue, there are strong structural issues that govern privatisation in Iran and these are likely to force it through regardless of which government comes to power (at least in the short term). This view was accepted by members of the elite during the interviews. As one interviewee put it: "The privatisation programme will survive. The society wishes for this to happen (Interview with Mr Ali Khorram (Former Ambassador to China), Deputy, High Council of Free Trade – Industrial Zones Secretariat, Tehran, 17 December 2005).

Analysis of Policymakers' Views

Although this study has tended to point out the differences between Iran and other states in order to tease out the effects of Iranian idiosyncrasies on privatisation, there are some similarities between Iran and other countries in this respect. In particular, bureaucratic influence seems to be playing a large part, as far as the elite can see. For instance, many interviewees made the point that one of the primary obstacles to privatisation has been the reticence of "middle managers and other bureaucrats" fearing they might lose their influence.

Similarly, there were some notable class differences in Iran which might have an effect on the future course of privatisation – as judged by the political, business and academic circles. Almost all interviewees recognised these differences, although not all of them thought that the differences would go in the same direction. For instance, some thought that the poor were ambivalent about privatisation, while others saw them as being very much against it. But, most believed that the political elite in general approved of privatisation.

Because the narrative in this chapter is based on field surveys, the findings are of special importance in the analysis of the privatisation process as they tell the story from the inside. The information derived from the fieldwork is set out as follows:

- It was widely accepted through experience that privatisation reduces the burden of the government.
- There is a mixed feeling about the privatisation process being carried out through the stock exchange and/or through limited tender/negotiation.
- Due to the restricting clauses in various sections of the constitutional laws relating to foreign investment, no significant progress was made in attracting foreign investment.
- Interest groups and pressure groups played a major role in decision-making and implementation.
- From the outset the privatisation process has been the subject of disputes and heated debates between the religious establishment and other influential bodies.

These structural factors needed to be taken account of in our analysis. In this respect, the data extracted from the interviews and the questionnaire have proved useful in the overall political, economic and theoretical analysis of privatisation. In the concluding chapters of this book the data thus obtained is placed in the broader context of my investigation of this subject and, along with other considerations presented in the preceding pages, much will have been acquired to answer the key question of what the determinants of privatisation in Iran are.

NOTES

1 The introduction to this section draws heavily on the work of Shahriari-Rad (2002).
2 The economy is based neither on the unlimited freedom of private ownership typical of capitalism, nor on the public ownership that denies private property rights (Taleghani, 1983).
3 Personal communication. Discussions with Mr Houshang Naderian, Chairman & Managing Director, Audit Organization, Tehran, December 2005.
4 Personal communication. Discussion with Dr Manouchelir Lighrani, M.D., former landowner, Tehran, December 2004.
5 This section is based on information provided by the Privatisation Organisation (2002).
6 Officially, the Iranian privatisation programme began in 1995, when the origins, the reasons for its popularity, requirements, management, social and economic consequences were relatively new to Iran and other countries in the Persian Gulf (Molawi, 2003).
7 One example among many is the power of the Minister of Economic Affairs and Finance who chairs the Assembly of the Privatisation Organisation, and also plays a crucial role in the appointment of the members of the Arbitration Commission, which is supposed to be independently adjudicating the contracts prepared and signed by the Privatisation Organisation.
8 The following section draws heavily on a report by the United States Information Administration (USEIA), 2005.
9 Personal communication. Discussions with Mr Hossein Kazempour Ardebili, Deputy Minister of Petroleum/Head of Iran delegation to OPEC, Member and Chairman of NIOC Board of Directors, December 2005.
10 It is unclear whether the National Assembly rejected the right of a company to develop its discovery. The company that made the discovery does not have an automatic right to exploit the discovery, but is given priority and, under certain conditions, can invest further and develop the field. Its share of production would be set according to the buy-back agreement.
11 It is reported that some Indian companies are also discussing involvement in this field.
12 Austrian Drilling Oil Company.
13 Personal communication. Discussion with Engineer Mohamed Hossein Nejad Hosseinian, Deputy Minister of Petroleum, International Affairs, Tehran, December 2005.
14 Although Saudi Arabia's oil production capacity was about 10 million bpd in the past, this is being increased to about 12. Currently, it is probably about 11 or 11.5. Saudi authorities have announced that they have plans to raise capacity to 15 million bpd, but have not yet started to implement the necessary field operations.
15 For instance, as noted by Dr Aghili, Managing Director, Karafarin Bank, Tehran, 29 December 2005.

6

CONCLUSION: PROSPECTS FOR PRIVATISATION

Summary of Arguments

The aim of this book is to examine the issues surrounding privatisation of some of Iran's national institutions, focusing on the oil and gas industries. By determining the critical factors which influenced policy-making, some understanding of the reasons behind the checkered course of privatisation in Iran from the Pahlavi regime to the present has been gained. More specifically, this study has investigated the determinants of the privatisation policy since the death of Khomeini to ascertain whether there has actually been the political will to privatise the Iranian oil and gas industries. To answer this question, various sub-themes were addressed, including the domestic and international environments over the past 60 years, along with the past privatisation policies of both Iran and other states.

The study started with an overview of the theoretical basis of privatisation by detailing different schools of thought. The central question was also examined by considering the opportunities for privatisation at a theoretical level and, concretely, the experience of other countries in privatisation, namely Pakistan, Azerbaijan, Kyrgyzstan and India, along with the non-neighbouring countries, Egypt and China. In this way, it has been possible to identify and appreciate the opportunities open to Iran for privatisation.

Through a politico-economic approach, Chapters 2, 3 and 4 examined the historical aspects which led to the present discussions of Iranian privatisation plans in general and that of the oil and gas industry in particular. Each chapter explored a specific era of Iran's history. Chapter 2 surveyed the post-Second World War period, dominated by the Shah, which ended with the Islamic Revolution in 1979. Chapter 3 then studied Khomeini's rule from 1979 to 1989. Finally, Chapter 4 addressed the contemporary situation in Iran since 1989. Each of these chapters explored in detail the political and economic features of the country, culminating in an analysis of the interplay between them. Having established the theoretical and historical contexts, this study proceeded to address more specifically the contemporary determinants of Iran's privatisation policy, and Chapter 5 then critically examined the four primary factors which directly affected Iran's privatisation policy.[1]

Summary of Conclusions

Taking all of the chapters together we can draw some conclusions about the determinants of the privatisation policy towards the oil and gas industry, among others, in Iran. These wide-ranging determinants are based on the multiple perspectives and methodologies used in this case study. The conclusions reached in each chapter are interwoven with the narrative, but for more precision they are consolidated and summarised here under subject headings.

Theory and precedents

Chapter 1 examined the theoretical boundaries and constraints on the determination of privatisation policy in Iran. In doing so, it effectively defined the landscape in which privatisation could take place.

It was found that there are at least four theoretical factors that must be taken into consideration in our analysis of privatisation of the Iranian oil and gas sector. These are:

- The definition and understanding of what privatisation is;
- the boundaries set by other forms of economic change and alternatives to privatisation;
- the different types and shades of privatisation;
- the limits and obligations within any privatisation transaction as set by the basic concept of the right to property.

Within these theoretical boundaries, the empirical information and historical background of privatisation were incorporated in each chapter. There were also some specific lessons to be learned from the different models of privatisation that were investigated, as well as from the historical precedents of privatisation. These lessons contributed directly to the decisionmaking process, showing that,

- even though privatisation may be agreed by the leadership, the nature of the method of privatisation will have a huge effect on how fast it is carried out, and this factor must therefore be considered in the analysis and predictions of future investment;

- it is often assumed that, in theory, the government will make fairly correct analyses of the benefits of any privatisation package and that the final policy decision will conform to them. However, one must keep an open mind about the possibility that such an assumption can be wrong, that the government can make mistakes;
- different industries may or may not be privatised depending on their type; specifically, so-called strategic industries, including the oil and natural gas industry, will tend to be privatised differently from other non-strategic industries;
- one must keep the possibility of other options open when considering what decisionmakers in Iran take into account (e.g. precedents) in their analysis of whether to privatise the oil and gas industry and how;
- political infighting over privatisation is the norm, not the exception. But even if the core of the Iranian government finally decides not to privatise the industry, there would have been an ongoing debate that might eventually produce a compromise.

The chapter has also revealed that the Iranian state-controlled economy has been unsuccessful in improving the lives of its people, as was promised by the founder of the Islamic Revolution and echoed to some extent by President Mahmoud Ahmadinejad.

Iran's political economy

The political economic analysis undertaken in the historical chapters (2, 3 and 4) demonstrated the truism that there is an interplay between domestic and international factors, be they political or economic. In the case of Iran, this has been particularly palpable in as much as it has determined the country's policy of privatisation. Each of these factors was explored separately.

In the domestic field, the first factor is the disjunction between Iran's government and society. This issue has been turned into a theory called "the theory of arbitrary rule" according to Katouzian (2000, 2003). This factor has traversed time and affected all governments in Iran, even after the revolution of 1979. The second factor is the disjunction between economic and political/social modernisation. This was examined in relation to the Shah's attempts at economic and political reform and thereafter up to this day. The third factor determining policy is Iran's dependence on oil. Oil played a major role in the Shah's Iran and has continued to do so since the revolution. It has skewed both domestic and international policy of the state up to the present. In the international sphere, the fourth factor that was considered is the relationship of Iran with the West (the United States in particular), and Iranian society's perception of this relationship, including the legacy of the Anglo-Iranian Oil Co and the coup against Dr Mossadeq. In particular, the perception of the West's imperialistic attitude towards Iran had significant effects on the nature of Iran's policy before the revolution and has continued to do so since.

These four factors have all helped to determine the nature of Iranian policy in regard to privatisation throughout the country's history, going back at least to the Shah's regime. They continued into the post-revolutionary period owing, in part, to the brevity of the Iranian revolution. Beyond these factors, there are two supplementary factors that have held sway since 1979. One is the role of the Islamic Republic's revolutionary ideology. After the revolution, politics and economics continued in the same vein to some extent, but huge changes took place since. The revolutionary ideology led to changes in the government (including factionalism) as well as in foreign policy. In regard to the latter, the hostage crisis comes to mind and, to a lesser degree, Iran's neutrality during the Gulf War of 1991. But the key issue today is Iran's nuclear development. The second supplementary factor is the role of religion. In particular, religion has had a complex, but very direct, bearing on privatisation. It has also had other effects on politics and economics owing to Khomeini's intricate views on revolution and fundamentalist Islam.

Perceptions of privatisation in Iran's society

Chapter 5 analysed, from a contemporary perspective, four of the primary factors affecting privatisation in Iran. It examined the background issues that led up to the current views on privatisation. It also studied the role of the legal and regulatory framework of privatisation. It concentrated attention on the unique importance of oil and natural gas to the country. Finally, it incorporated the primary fieldwork information derived from the personal views of Iranian policymakers and those who carry out policy. The general conclusions reached here regarding the legal, regulatory and oil issues are straightforward. Legal and regulatory issues are, at the broadest level, tools of policy. While there are uncertainties about the effects of the legal and regulatory policies already on the books, there are even greater uncertainties about the immense role of oil and that of the elite in the economic life of Iran.

Because of its strategic importance, oil occupies a special place in terms of privatisation. Therefore, it is not appropriate to consider oil and gas privatisation in the same way as that of other industries. This being so, it was necessary in this study to take a more direct look at some of the specific concerns of the elite in Iran.

As was alluded to in the sections dealing with the political economy in theoretical terms, there is a significant difference between those in government or those connected to its higher echelons (the elite) and the common people. Since members of the elite are the ones who are actually forming policy, it was helpful to weigh their perception of privatisation in Iran and also to learn how the common people judge it, if at all. The difference between these groups is well summarised by one interviewee in particular, who noted:

> The blue-collar workers and the farmers had no clue as to what was and is going on. The higher income and middle management as well as the political elite and policy-makers think very highly of the privatisation scheme (Gholam Hassan Abiri, Lecturer, MPO and Tehran University, Applied Economics Research, Tehran, 19 December 2005).

Though there is consensus among members of the elite that further privatisation is bound to happen, some are more concerned than others about the pace at which it would proceed, considering the systemic delays in decisionmaking and execution; they also wonder about which parts of the oil and gas industry would be earmarked for privatisation and in what order. However, despite the change of government in 2005 and the deteriorating foreign relations, the widespread feeling is that more privatisation would take place all the same because, come what may, the structural deficiencies in the economy would require it.

Factors Determining Privatisation Policy

Table 6.1 encapsulates the factors determining privatisation discussed in the preceding chapters.

TABLE 6.1
Factors Determining Iran's Privatisation Policy

	Factors Determining Privatisation					
Historical and Political Economy	Disjunction between state and society	Disjunction between economic modernisation and political structure	Economic dependence on oil	Relations with, and perceptions of, Iran's relations with the West, especially the USA	Revolutionary ideology	Religion
Theoretical	Theories of privatisation	Alternatives to privatisation				
Precedents	Neighbouring states: Pakistan, Azerbaijan, Kyrgyzstan and India	Other important states: Egypt and China				
Specific	Background to privatisation	Legal and institutional factors	Oil sector	Personal perceptions of policymakers		

Source: Privatisation Organisation (2004).

Thus, we are able to conclude that the answer to the question raised in this study is this: privatisation policy in Iran will be determined by several interacting factors including the disjunction between state and society, the disjunction between economic modernisation and the political structure, Iran's economic dependence on oil, relations and perceptions of relations between Iran and the West (the United States in particular), the Islamic revolutionary ideology and religion, as well as theories of privatisation and international precedents of privatisation. However, based on a reading of the themes that have emerged throughout this study, it is possible to make a more inclusive and conceptual conclusion. The two key themes relate to structural factors and the notion that Iran is an exceptional case.

Iran's "exceptionalism"

This study has found, in both its primary and secondary research, the persistent theme of what one may call Iranian "exceptionalism". Although many states have come to privatise their national industries, it seems that the case of Iran is somehow exceptional. This view permeates through much of the literature on Iran and its economy, and is further illustrated in this study. The particular factors leading to this view, which were addressed throughout this study, come down to the role of religion and revolution in Iran.

Religion creates exceptionalism in regard to privatisation simply because Iran's system is an Islamic theocracy. This has meant that all politics and economics are affected to some degree by Islam, in particular, by the dictates of Shi'a Islam (Chapter 4). However, there are other Islamic states, and even ones that are also dependent on oil. Take Saudi Arabia, for instance. Iran differs from these other states in that it is not only a theocratic state but also a revolutionary one.

The Iranian revolution and its aftermath had an immense and, to a certain extent, unique impact on Iran. For example, as discussed earlier, the revolution led to foreign relations crises such as the hostage incident. At the same time, the revolutionary momentum has persisted to this day, influencing current political and economic decisions. Together, these factors (religion and revolution) have led to the exceptionalism of the country and Iranian privatisation policy more specifically. As noted, academics both within and outside Iran share this view of Iran's uniqueness. It was also prominent in the interviews and questionnaire responses.

For this reason, one should not take it to be the case that one can understand privatisation in Iran purely in terms of theory or practice – the experience of other countries. These were of some help in contextualising Iranian privatisation. In Iran's case the difference, of course, arises out of the fact that Iranian policies and economics are deeply affected by religion and revolution. As such, it was necessary to concentrate specifically on Iran's case and so this study took account of many issues that might not be relevant to examples of privatisation elsewhere.

Although this work has tended to point out the differences between Iran and other states to show up the effects of Iranian idiosyncrasies vis-à-vis privatisation, there are some similarities between Iran and other countries in terms of structural constraints on privatisation. In particular, bureaucratic power seems to be playing a large role, as seen from the elite's viewpoint. For instance, many interviewees made the point that one of the primary obstacles to privatisation has been middle managers and other bureaucrats whose fear of losing their influence makes them distinctly reticent in matters concerning privatisation.

It has been argued that Iranian exceptionalism is the result of a combination of the effects of religion, revolution and bureaucracy. On this reasoning, it has been suggested that the Iranian five-year development plans reflect some aspects of this combination. Iran's five-year central planning, first established in 1979, appears as the coalescence of religion and revolution in the form of Islamic Marxism. Over time, the effects of central planning and the views on it have varied. When analysing the outcome of the Third Development Plan covering 2000–2004, the most recently completed plan, Amuzegar says:

> Five-year central planning has now become a sacred ritual in the Islamic Republic . . . and although [it] has proven to be a costly exercise in futility, it is still regarded as a talisman for Iran's economic salvation (Amuzegar, 2005:61).

Djavad Salehi-Isfahani, Professor of Economics at Virginia Tech, suggests in turn that planning has become more indicative than centralist and that it fulfils a necessary function of coordination between the private sector and the government (Samii and Aman, 2005).

It ought to be said that Amuzegar is not entirely negative about the outcomes of the Third Plan either. He claims that these were much better than those of the previous plan (the Second Development Plan), yet he insists in pointing out the lack of effective formulation of development goals and the lack of meticulous implementation that seemingly characterise the Third Plan. It is suggested that both kinds of deficiencies are typical features of vast and largely ideologised state machinery like the Iranian administration. Furthermore, the lack of commitment among the different levels of bureaucrats to conform to the plan mandates and, at the same time, the unrealistic nature of some of the goals are also emphasised by Amuzegar as contributing factors to the partial failure of the Third Plan. He also complains that no real progress has been made in price decontrols and in privatising money-losing state enterprises under the Third Plan (2005:61).

The successor of the Third Plan, the Fourth Plan, was approved in September 2004 and contains six parts (Privatisation Organisation, 2004). The first part, which delineates 57 articles aiming at the development of a "knowledge-based growth of the national economy in interaction with the global economy", along with the sixth part, devoted to "good governance and the modernisation of the state", appear to be the most relevant ones to privatisation (ibid.). Beyond the more obvious concern with rapid economic expansion, there is clearly an interest in addressing the issues pointed out by Amuzegar (2005) in relation to the current inefficiency of the state apparatus (by improving governance and modernising the functioning of the public administration) which have direct implications for privatisation. However, the Fourth Plan had a shaky start. According to some legislators, the by-laws and instructions required to commence the implementation of the plan had not been prepared in time and had to be designed "on the march" (Samii and Aman, 2005). This kind of amending dynamic, that is the practice of designing the implementation measures as the central plans unfold – instead of planning ahead – seems to have become the rule within the Iranian administration. For all these reasons, it is suggested that the notion of exceptionalism is applicable in the context of government central planning, and that the five-year development plans actually reflect some key aspects of the triad that sustains the notion: religion, revolution, and bureaucracy.

There are also some perceivable class differences in Iran which seem to have an effect on the future of privatisation as seen by the political, business and academic elite. Almost all interviewees noted these differences, although not all thought that the differences would lead to the same results. For instance, some thought that the poor were ambivalent about privatisation, while others saw them as being very much against it, unlike those who said they had no clue. Anyway, most believed that the political elite in general was in favour of privatisation.

Nevertheless, while these opinions have played a role, they do not override the importance of religion and revolution. This is illustrated by the fact that Iran did not privatise its oil and gas industry in the same way or as fully as other countries had done; they had no exceptional issues to deal with.

Structural factors

There are also structural factors that need to be taken into account for an understanding of the privatisation of oil and gas in Iran. Gidden's (1984) concept of structuration has already been discussed. There are multiple structural factors that have had, and will continue to have, an overwhelming causal effect on privatisation in Iran. These include

- the gradual shift of Iranian internal and foreign policy towards a more pragmatic approach, which is in response to the realities of internal politics and foreign affairs;
- the growth and shrinkage of the Iranian economy, which are governed by oil prices;
- the gradual trend of inefficiencies in the Iranian economy as the nationalised oil industry has evolved over time and in relation to the regional and global economy;
- the evolution of internal Iranian politics, in particular, in regard to the high proportion of young people and a more educated population which is more connected to the outside world;
- the exceptionally prominant role of religion and revolution and their evolution over time, which in this sense also count as structural factors.

Together, these factors are more fundamental than any particular policy or the attitude of any politician. They are also more relevant than any particular external foreign policy or political event, including the 2003 war in Iraq or the ongoing nuclear crisis. Thus, privatisation in Iran must be understood within the context of an ongoing structural progression. One should conclude that these structural factors have been inextricably pushing Iran towards privatisation. It is clear at an empirical level that Iran needs to deal with inefficiencies in the economy. It is also the case that Iranian privatisation policy has continued to move forward, though at a variable pace, for over a decade. This was made clear by the views of members of the elite whom I interviewed for this study.

Factors Determining Future Prospects

Such a conclusion should not be taken too far. Although this study finds that a more comprehensive privatisation is bound to happen in Iran, it is less clear as to how and when. Three factors must be taken into account in order to make more precise conclusions and find more precise implications. These are the unpredictability in internal politics, the effects of the global economy, and the differences inherent in Iran's oil and gas economy, especially in regard to upstream and downstream privatisation.

The main issue for Iran at the moment is its stand-off with the international community over the nuclear issue. What is obvious, is that the stand-off is in sharp contrast to Khamenei's declaration of 3 July 2006 to the effect that the resolution of this crisis will impact favourably on the future course of Iran, not just in terms of its foreign policy, but also in terms of its domestic politics, its economy and potentially even the nature of the state itself.

The problem of unpredictability

An important issue covered in Chapter 5, which I undertook to return to, is the problem of unpredictability in national politics. In particular, it was noted that the election of Ahmadinejad was described by some analysts as unexpected. And, pertinently to this discussion, the particular policies and policy paths which he has chosen are not necessarily those which accord with the point made earlier, namely that there has recently been a reversal of the policy to modernise, thereby causing privatisation to slow down. Even with the significant amount of information that this study has provided, it would be foolhardy to make firm predictions.

At least we can say for certain that the pace and extent of privatisation will be determined to some degree by changes in Iranian politics. Although we have come to the clear conclusion that further privatisation will happen regardless, it must be admitted that changes in the leadership of Iran (such as the election of Ahmadinejad four years ago) do have an effect on the *pace* of privatisation. There is little doubt that had Rafsanjani won the election in 2005 the pace and extent of privatisation would have been greater.

The effect of globalisation on Iran

This study has frequently brought out the effects of the regional as well as the global politico-economic order on Iran. This was experienced by Iran decades ago when the United States covertly organised a coup against Mossadeq and reinstated the Shah for *economic reasons*. Such a rationale continues to this day. Even though Iran isolated itself, or was isolated by many foreign states after the Islamic Revolution, globalisation has meant that it could not escape the effects of the global economy and the foreign policies of other states. Iran's view of globalisation is that it is a tool in the hands of the former imperialists and the current superpower to legitimise the neo-liberal global project, that is the creation of a global free market and the consolidation of Anglo-American capitalism around the world (Held and McGrew, 2002). In post-revolutionary Iran, there has been no sustained dialogue between the "globalisers" and the Iranian leadership. The revolution swept away the secular, pro-Western elite and delivered the state to Islamists, thus removing the possibility of harmonising with globalisation. Instead, the Islamic state turned in on itself, both because of the imposition of sanctions by the United States and because of the perceived need to consolidate and defend the revolution (Henry and Springborg, 2001). Notwithstanding this phenomenon, I contend that Iran has been and will continue to be sensitive to global issues, since the elasticity of its oil revenues is linked to its awareness of, and connection to, the global systems.

A particularly significant issue to take account of is the role of world oil prices. As was noted earlier in this study, changes in oil prices affect the prospects of Iranian privatisation of the oil and gas industry. In simple terms, when oil prices are relatively high, progress in privatisation tends to slow down; when oil prices are relatively low, it tends to speed up.

The logic behind this is fairly straightforward: when oil prices rise, its worth to the Iranian government rises and therefore it is less willing to part with it for the sake of privatisation. Since the government has more money at such times, it is able to make up for inefficiencies in the economy and support itself, remain politically stable and, in general, prop up the overall economy. This being so, it does not need to reform and privatise the oil and gas industry. The logic goes in reverse as well: when oil prices fall, the government has less money to prop itself up, maintain the overall economy and the level of subsidies. This means that it is forced to look for ways to make the economy more efficient, and privatisation is an obvious way to do this.

Upstream versus downstream oil privatisation

Chapter 5 referred to a discernable difference in the prospects of upstream versus downstream privatisation in Iran. This point needs to be further reinforced. In particular, we need to take into account three factors: oil and gas, which are considered to be strategic industries in the country, and an Iranian exceptionalism.

The oil economy in Iran is strategic on multiple levels. At one level it is necessary in order to keep the total Iranian economy afloat since its economy is very much dependent on oil as an export that brings in foreign capital and as a means of employing (or subsidising) a large number of people in Iran. At the same time, Iran's oil and gas give the state significant strategic weight in its dealings with neighbouring states and elsewhere across the world in that it can threaten to lower or even stop oil and gas exports. Finally, oil and gas are used generally as a means of strategic trade as illustrated by recent transactions between Iran and China. Together, these factors mean that the Iranian government cannot easily let go of its control over its oil- and gas-based economy and put its fate in the hands of companies domiciled in countries that may potentially become enemies.

The strategic value of Iran's oil and gas industry is compounded by Iranian exceptionalism. As noted, Iran's politics and economics are dictated to some degree by religious and revolutionary undertones. In this respect, other states (and even companies from other states) are seen through the lens of revolution and (different) religion. This means that Iran is seen as less trustworthy as regards international agreements, economic or otherwise, than other countries.

All these factors have bred a reluctance on the part of the government to privatise the oil and gas industry. Hence the reluctance to privatise some parts of the oil and gas industry compared to other parts of the industrial sector. In particular, Iran has been less willing to privatise those areas of the oil and gas industry which are more geared to domestic consumption than others, for fear of becoming dependent on other states. In short, it is less willing to privatise the upstream oil and gas industry than that located downstream.

Implications and Prospects

We can now envisage the prospects of privatisation in the future from the conclusions of this study. These prospects can be summarised as follows: privatisation will continue but its pace will remain dependent on internal politics and global oil prices; the privatisation of the downstream oil and gas industry will continue, to be followed by that of the upstream industry at some time in the future. As indicated earlier an important conclusion of this study is that further privatisation is bound to take place in Iran and will eventually include the privatisation of most of the oil and gas industry. This is demonstrated by the structural factors of Iran's political and economic system, independent of short-term political or economic changes. Rather, the question of privatisation in Iran is a matter of pace and extent.

The pace of Iran's privatisation of the oil and gas industry will depend on multiple factors, the most important of which are changes in the internal political situation and changes in global oil prices. Internal political changes, such as the continued governance (or re-election) of Ahmadinejad, will merely lead to a slowdown in the pace of privatisation. On the other hand, the pace of privatisation will also depend on changes in world oil prices. The prospect of continued high world oil prices lead us to conclude that privatisation will proceed at a slower rate in Iran. In any event, for the privatisation of oil and gas to succeed in Iran, the state organisation will need to be restructured. The prerequisites are:

- National financial management must be entrusted to professionals and not relegated to others on the basis of their political affiliations.
- At present, the country lacks a clear and comprehensive investment policy, as well as the rules and regulations governing investment, especially foreign investment. The government should engage independent professionals from both within and outside the country to review the existing rules and regulations.
- There is a need to overhaul the bureaucratic system now in place. Independent financial consultants should be invited to provide objective solutions to the process of investment by local and foreign investors.
- Tax and labour laws have to be overhauled, enabling investors to make short-, medium- and long-term plans for their investments. Furthermore, tax and labour laws should be designed to create a favourable environment so as to attract foreign investment in capital-intensive projects.

- New regulatory and watch-dog organisations should be set up to monitor the performance of the privatised institutions.

To conclude, the extent of privatisation in Iran will depend on the strategic importance of the industry in relation to the exceptional factors of Iran's political and economic system. This means that one can expect a fast pace of privatisation to continue in non-strategic sectors such as farming. Within the oil and gas industry, one can expect that downstream industries to be privatised faster than upstream industries. Eventually one can expect both upstream and downstream industries to become fully privatised.

Although Ahmadinejad was elected to fight corruption, it is hard to imagine that the regime will allow him to do anything meaningful, because it is the regime itself that is the principal generator and beneficiary of corruption. It seems also highly unlikely that he will be any more effective in alleviating the plight of the Iranian people than President Khatami before him. While Ahmadinejad campaigned on a platform of economic reform, his vision of it echoes the populist version of 1970s socialism. In truth, Iran needs less socialism, not more. It needs fewer price controls, more privatisation, more private investment and more free enterprise.

NOTE

1 The four being the specific background of privatisation thought in Iran; the legal and institutional framework; the nature of oil and gas relative to privatisation; and the personal thoughts of the policymakers concerned.

APPENDIX 1

LIST OF INTERVIEWEES

Mr Gholam Hassan Abiri, Lecturer, MPO and Tehran University, Applied Economics Research, Tehran, 19 December 2005.

Professor Hossein Afarideh, MP, Member of the Committee for Energy at Majlis, Deputy to Iran's Head of Nuclear Agency, 24 December 2006.

Dr Parviz Aghili, Managing Director, Karafarin Bank, Tehran, 29 December 2005.

Mr Hossein Kazempour Ardebili, Deputy Minister of Petroleum/Head of Iran delegation to OPEC, Member of NIOC Board of Directors and Chairman of NICO. Interview took place at the ministry HQ on 26 December 2005 (also called into the meeting: Mr Mohammad Alipour-Jeddi, Head Petroleum Market Analysis Dept., OPEC, Austria.

Dr Mehdi Assali, President, Institute for International Energy Studies (IIES), Tehran, 19 December 2005.

Dr Mehdi M. Behkish, Secretary General, Int'l Chamber of Commerce-Iran, and Associate Professor in Economics, 18 December 2005.

Dr Bijan Bidabadi, Faculty Member & Director of Foreign Exchange Research Department of Central Bank of Iran, 28 December 2005.

Engineer Kamal Daneshyar, MP, Head of Energy Commission, Majlis, 24 December 2005.

Dr H. Ghanimi Fard, Executive Director, International Affairs, NIOC Petroleum Ministry HQ, Tehran, 22 December 2005.

Dr Narsi Ghorban, Managing Director, NarKangan (NGTLC) Gas to Liquid Int'l Co., Tehran, 18 December 2005.

Dr Ali Hamadani, Legal Advisor, Society for Industrial Management, Tehran, 17 December 2005.

Engineer Mohamed Hossein Nejad Hosseinian, Deputy Minister of Petroleum, International Affairs, Tehran, 18 December 2005.

Mr Ali Khorram (former Ambassador to China), Deputy, High Council of Free Trade – Industrial Zones Secretariat, Tehran, 17 December 2005.

Mr Hassan Khoshpour, Deputy Economic Affairs/Privatisation, Management and Planning Organisation (MPO), 26 December 2005.

Dr Akbar Komijani, Vice Governor, Economic Directorate, The Central Bank of Iran, 25 December 2005.

Mr Hassan Lajevardi, Lecturer and Deputy Director Econ. Research, Centre for Energy Studies-MATN, Tehran, 20 December 2005.

Mr Saeed Leylaz, Vice President-Sales and Marketing, Iran Khodro Diesel Co. Plc., 19 December 2005 at their HQ.

Dr Abbas Maleki, Chairman, International Institute for Caspian Studies, Tehran, 24 December 2005.

Mr Hossein Mamdooh, Member of Board of Directors, Iran Helicopter Services Company (IHC) and Head of Quality Assurance, Tehran, 22 December 2005.

Dr Alinaghi Mashayekhi, Economics Dept., Sharif Technical University, 24 December 2005.

Mr Houshang Naderian, Chairman & Managing Director, Audit Organization, Tehran, 22 December 2005.

Mr Iraj Nadimi, MP, Deputy Head of Majlis Economic Committee, and University Lecturer, Tehran, 24 December 2005.

Dr Massoud Nily, Head of Economics Dept., Sharif University, 31 December 2005.

Dr Ebrahim Razaghi, Professor of Economics, Tehran University, Tehran, 26 December 2005.

Dr Mehdi Sadeghi, Alameh University, Economics Dept., 28 December 2005.

Dr Sohrab Shahabi, Deputy Secretary General, Economic Cooperation Organization and University Lecturer (interview conducted in Tehran), 27 December 2005.

APPENDIX 2

LIST OF RESPONDENTS TO THE QUESTIONNAIRES

Dr Parviz Aghili, Managing Director, Karafarin Bank (first private bank in Iran), Questionnaire form was later used in the personal semi-structured interview of 29 December 2005.

Dr Abolghassem Hashemi, Associate Professor of Economics, Dept. of Economics, Shahid Beheshti University, 30 December 2005 (this form was filled out during a personal interview).

Mr Hossein Mamdooh, Member of Board of Directors, Iran Helicopter Services Company (IHC), a company owned by NIOC and singled out for privatisation, and Head of Quality Assurance, December 2005.

Mr Houshang Naderian, Chairman & Managing Director, Audit Organization, December 2005, very kindly filled out the questionnaire whilst also agreeing that I carry out the semi-structured interview with him as well (was appointed by the government to check the accounts and audit approximately 2000 government companies).

BIBLIOGRAPHY

Works Cited

Abrahamian, E. (1982). *Iran between Two Revolutions*, Princeton, New Jersey, Princeton University Press.

Abrahamian, E. (1991). 'Khomeini: fundamentalist or populist?', *New Left Review*, 186, 102–9.

Abrahamian, E. (1993). *Khomeinism: Essays on the Islamic Republic*, Berkeley, University of California Press.

Abu Shair, O.J.A.R. (1997). *Privatization and Development*, Basingstoke, Macmillan.

AFP (2006). 'Iran unveils major privatisation plan', 3 July, Agence France Press, available on-line at http://forum.greens.org/pipermail/grns-gpoc/2006-July/000300.html.

Ahmad, K. (ed.) (1980). *Studies in Islamic Economics: A Selection of Papers Presented to the First International Conference on Islamic Economics Held at Makka under the Auspices of King Abdul Aziz University, Jeddah, February 21–6, 1976*, Jeddah, International Center for Research in Islamic Economics.

Ahmad, Z. (1991). *Islam, Poverty and Income Distribution*, Leicester, The Islamic Foundation.

Ahmed, E. (1996). *Role of Private and Public Sector in Economic Development in an Islamic Perspective*, Herndon, International Institute of Islamic Thought.

Ahsan, M. (1991). 'Population administration during the time of Prophet Mohammad', *We People* (biannual journal of the government of Pakistan), VI (July–December), 15–16.

Ala, M. (2000). 'Privatisation of the National Iranian Company', paper presented at the 75th BIBA Business Meeting, 10 May.

Alizadeh, P. (ed.) (2000). *The Economy of Iran: Dilemmas of an Islamic State*, London, I. B. Tauris.

Amini, A. (2004). *Oil: Politics and Economics*, Tehran, Third Line (Khate Sevom).

Amuzegar, J. (1999). *Managing The Oil Wealth, OPEC's Windfalls and Pitfalls*, London, I. B. Tauris.

Amuzegar, J. (1997). *Iran's Economy Under the Islamic Republic*, London, I. B. Tauris.

Amuzegar, J. (2005). 'Iran's third development plan', *Middle East Policy*, XII (3), 61.

Amuzegar, J. (2007). 'Iran's new energy picture', *Middle East Economic Survey*, 50, 5.

Anon (1997). Special Issue on Leaders and Leadership, *International Journal*, LII (2), Spring.

Anon (2004a). *Business Weekly*, 10 December.

Anon (2004b). 'Iran: the state of Iran', 15 January, available on-line at http://www.economist.com/displaystory.cfm?story_id=S%27%298%20%29PQ%3B%2A%210%22%3C%0A&tranMode=none.

Anon (2005a). *Economist*, 25 November.

Anon (2005b). 'Iran: political forces', 5 May, from the Economist Intelligence Unit, available on-line at http://www. economist.com/countries/Iran/profile.cfm?folder= Profile%2DPolitical%20forces.

Anon (2005c). 'Ahmadinejad embarrassed again', from *The Economist Global Agenda*, 23 November, http://www.economist.com/agenda/displaystory.cfm?story_id=5212878.

Ansari, A.M. (2000). *Iran, Islam and Democracy. The Politics of Managing Change*, London, The Royal Institute of International Affairs.

Arab-British Chamber of Commerce (1990). *Privatisation. Principles and Practice. A Survey of Privatisation Policies in the Industrial, Developing and Arab Countries*, London, Arab-British Chamber of Commerce.

Azzam, H.T. (1995). *Privatization in the Gulf Countries*, United Nations Economic and Social Council, Economic and Social Commission for Western Asia.

Backhaus, J. G. (1989). 'Privatisation and nationalisation – a suggested approach', *Annals of Public and Cooperative Economics*, 3, 307–8.

Bamberg, J. H. (1994). *The History of the British Petroleum Company – The Anglo-Iranian Years 1928–1954*, Cambridge, Cambridge University Press.

Banerjee, S.G. & Munger, M.C. (2004). 'Move to markets? An empirical analysis of privatization in developing countries', *Journal of International Development*, 16, 213–40.

Barzel Y. (1989). *Economic Analysis of Property Rights*, Cambridge, Cambridge University Press.

Basu, P. K. (1994). 'Demystifying privatization in developing countries', *International Journal of Public Sector Management*, 7 (3), 44–55.

BBC (2004a). 'Iranian MPs back investment veto', 22 September, available on-line at http://mews.bbc.co.uk/1/hi/business/3680056.stm.

BBC (2005a). 'Iran: who holds the power?', London, BBC.

BBC (2005b). 'Country profile: Iran', report from BBC 5 o'clock News, 3 November, also available on-line at http://news.bbc.co.uk/go/pr/fr/-/1/hi/world/middle_east/country_profiles/790877.stm.

BBC (2005c). 'Iran's new leader: a familiar face?', reporter John Simpson, BBC world affairs editor, report from BBC News, 31 August, also available at http://news.bbc.co.uk /go/pr/fr/-/1/hi/ world/middle_east/ 4626081.stm.

BBC (2005d). 'Iran's new man - in his own words', report from BBC News, 27 June, also available on-line at http://news.bbc.co.uk/go/pr/fr/-/ 1/hi/world/middle_east/ 4626479.stm.

BBC (2006). 'Profile: Mahmoud Ahmadinejad', report from BBC News, 28 April, also available on-line at http://news.bbc.co.uk/go/pr/fr/-/ 1/hi/world/middle_east/ 4107270.stm.

Beesley, M. E. & Littlechild, S. C. (1989). 'The regulation of privatised monopolies in the United Kingdom', *Rand Journal of Economics*.

Behdad, S. (2000). 'From populism to economic liberalism: the Iranian predicament'. In: Alizadeh, P. (ed.) (2000). *The Economy of Iran: Dilemmas of an Islamic State*, London, I. B. Tauris.

Betz, H. K. (1988). 'How does the German historical school fit', *History of Political Economy*, 20 (fall), 409–30.

Bill, J. A. & Springborg, R. (1994). *Politics in the Middle East*, New York, Harper Collins.

Bird, G. (1992). *Economic Reform in Eastern Europe*, Aldershot, Edward Elgar.

Bishop, M., Kay, J. & Mayer, C. (1994). *Privatization and Economic Performance*, Oxford, Oxford University Press.

Blanchard, O., Dornbusch, R., Krugman, P. R. & Layard, R. (1991). *Reform in Eastern Europe*, Cambridge, Massachusetts, MIT Press.

Bouin O. (1992). *Privatisation in Developing Countries: Reflections on a Panacea*, Paris, OECD Development Centre, Policy Brief No. 3.

Buchta, W. (2000). *Who Rules Iran? The Structure of Power in the Islamic Republic*, Washington DC, The Washington Institute for Near East Policy.

Butler, E. (ed.) (1992). *Privatization in the Nineties*, London, Adam Smith Institute.

Byman, D.L. (2005). 'Understanding Iran's nuclear agenda', *The Chronicle of Higher Education*, available on-line at http://chronicle.com/weekly/v51/i37/37b01201.htm.

Chandra, R. (1992). *Industrialization and Development in the Third World*, London, Routledge.

Chapra, M.U. (1992). *Islam and the Economic Challenges*, Leicester, The Islamic Foundation and the International Institute of Islamic Thought.

Choudhury, M. A. (1998). *Studies in Islamic Social Sciences*, London, Macmillan.

Clawson, P. (2003). 'Impact of World Bank Loans to Iran', 29 October. Testimony before the House Committee on Financial Services, Subcommittee on Domestic and International Monetary Policy, Trade, and Technology, available on-line at http://www.washingtoninstitute.org/print.php.

Clist, M. (2006). 'World energy outlook. Investment in Iran. A must', *Economic Trends*, 3(1), SE.

CNN (2003). 'World News', broadcast 3 June, quoting President Bush at the Red Sea Summit.

Cook, P. & Kirkpatrick, C. (1995). *Privatisation Policy and Performance, International Perspectives*, Hertfordshire, Prentice Hall/Harvester Wheatsheaf.

Coville, T. (2005). 'The Iranian economy, a new start or a continuation?', 23 October, available on-line at http://www.caucaz.com/home_eng/breve_contenu.php?id=198.

Dana (2004). Company newsletter, Dana Insurance Company Limited, Tehran.

Dawkins, D. (ed.) (1991). *Power and Politics in Education*, London, Falmer.

Demsetz, H. (1967). 'Towards a theory of property rights', *The American Economic Review – Papers and Proceedings*, 2, 347–59.

Drollas, L. & Greenman, J. (1989). *Oil, The Devil's Gold*, London, Gerald Duckworth & Co.

Economist Intelligence Unit (2004). 'Economic structure', 14 April, from the Economist Intelligence Unit, available on-line at http://www.economist.com/countries/Iran/profile.cfm?folder=Profile%2D Economic%20Structure.

Economist Intelligence Unit (2005). 'Political forces', 5 May, from the Economist Intelligence Unit, available on-line at http://www.economist.com/countries/Iran/profile.cfm?folder=Profile%2D political%20Forces.

Economist Intelligence Unit (2006). *Country Report.*

ECSSR (The Emirates Center for Strategic Studies and Research) (1999). *Privatization and Deregulation in the Gulf Energy Sector*, Abu Dhabi, The Emirates Center for Strategic Studies and Research.

Ehteshami, A. (1995). *After Khomeini, The Iranian Second Republic*, London, Routledge.

Elm, M. (1994). *Oil, power and Principle – Iran's Oil Nationalization and its Aftermath*, Syracuse, Syracuse University Press.

Encyclopaedia of the Orient (2004). *Encyclopaedia of the Orient*, available on-line at www.Lexicorient.com.

Ferrier, R. W. (1982). *The History of the British Petroleum Company*, Cambridge, Cambridge University Press.

First Initiative (2005). 'Information Exchange: Iran', 25 February, available on-line at http://www.firstinitiative.org/InformationExchange/Countries/country_information.cfm?iEligibleCountryID=64.

Fuller, G. E. (1991). *The Center of the Universe, the Geopolitics of Iran*, A Rand Corporation Research Study, Boulder, Colorado, Westview Press.

Ghobadian, A. & Viney, H. (2000). 'Issues in strategic investment in the global electric utility industry', *SAM Advanced Management Journal*, Autumn.

Ghobadian, A., James, P., Liu, J. & Viney, H. (1998). 'Evaluating the applicability of the Miles and Snow typology in a regulated public utility environment', *British Journal of Management*, 9(S1), 71–83.

Giddens, A. (1984). *The Constitution of Society: Outline of the Theory of Structuration*, Berkeley, University of California Press.

Globalsecurity (2005). Information available on-line at Globalsecurity.com.

Goldman, M. I. (2003). *The Piratization of Russia, Russian Reform Goes Awry*, London, Routledge.

Gormley, W. Jr. (ed.) (1991). *Privatisation and its Alternatives*, Wisconsin, University of Wisconsin Press.

Gourevitch, P. (1978). 'The second image reversed: the international sources of domestic politics', *International Organization*, 32(4), 881–912.

Hakiman, H. & Karshenas, M. (2000). 'Dilemmas and prospects for economic reform and reconstruction in Iran'. In: Alizadeh, P. (ed.) (2000). *The Economy of Iran: Dilemmas of an Islamic State*, London, I. B. Tauris.

Halliday, F. (2003). *Islam and the Myth of Confrontation*, (2nd edn.), London, I. B. Tauris.

Hanousek, J. & Kocenda, E. (2003). 'The impact of Czech mass privatisation on corporate governance', *Journal of Economic Studies*, 30, 3–4.

Hoopes, S. M. (1997). *Oil Privatization, Public Choice and International Forces*, London, Macmillan.

Hossain, M. & Malbon, J. (eds.) (1998). *Who Benefits From Privatisation?*, London, Routledge.

Ibrahim, F. & Ibrahim, B. (2003). *Egypt: an Economic Geography*, New York, I. B. Tauris.

IMF (1995). *Country Report: Azerbaijan Republic: Recent Economic Development*, Washington, International Monetary Fund.

IMF (2003). *Islamic Republic of Iran: Staff Report for the 2003 Article IV Consultation*, 1 August, Washington, International Monetary Fund, p. 34.

Iran Daily (2004). 'Modern Banking Strategy', 28 November, available on-line at http://www.iran-daily.com/1383/2150/html/focus.htm.

Iranchamber (2005). 'Ayatollah Seyed Ali Khamenei'. Details available on-line at http://www.iranchamber.com/history/akhamenei/ali_khamenei.php.

Jervis, R. (1976). *Perception and Misperception in International Politics*, Princeton, Princeton University Press.

Jervis, R., Lebow, R. N., Stein, J. G., Morgan, P. M. & Snyder, J. L. (1985). *Psychology and Deterrence*, Baltimore, Johns Hopkins University Press.

Kamrava, M. (1990). *Revolution in Iran: The Roots of Turmoil*, Routledge, London.

Kanovsky, E. (1997a). *Iran's Economic Morass: Mismanagement and Decline under the Islamic Republic*, Policy Paper No. 44, Washington DC, The Washington Institute for Near East Policy.

Kanovsky, E. (1997b). *Recent Economic Developments in the Middle East*, Tel Aviv, Shiloah Center for Middle Eastern & African Studies, Tel Aviv University.

Katouzian, H. (1981). *The Political Economy of Modern Iran 1926–1979*, London, Macmillan.

Katouzian, H. (1989). 'The political economy of Iran since the revolution: a macro-historical analysis', *Comparative Economic Studies*, 31(3), 55–66.

Katouzian, H. (2000). *State and Society in Iran: The Eclipse of the Qajars and the Emergence of the Pahlavis*, London, I. B. Tauris.

Katouzian, H. (2003). *Iranian History and Politics: the Dialectic of State and Society*, London, Routledge Curzon.

Keynes, J. M. (1936). *The General Theory of Employment, Interest and Money*, London, Macmillan.

Khalili, M. & Manafi, S. (2000). *Fundamentals of the Islamic Revolution*, Tehran, The Institute for Compilation and Publication of Imam Khomeini's Works.

Khavand, F. (1999). *An Overview of Iranian Economy after Revolution*, cited in *Iran Nameh*, Iranian Study Foundation, Bethesda, USA, pp. 645–56.

Khazaee, M. (2003). 'Foreign investment in Iran: advantages and opportunities', speaking on 30 June at Royal Institute of International Affairs conference entitled 'Iran', Royal Institute of International Affairs, Chatham House, London, 30 June – 2 July.

Klare, M. (2005). *Oil, Geopolitics, and the Coming War with Iran*, New York, Mother Jones.

Knight, F. (1950). *Introduction to Carl Menger's Principles of Economics* (translated and edited by James Dingwall and Bert Hoselitz), New York, Free Press.

Koussari, S. (1999). *The Iranian Oil and Gas Report. The International Guide to the Iranian Oil and Gas Industry*, London, SMI Publishing.

Krasner, S. D. (1978). *Defending the National Interest: Raw Materials Investment and U.S. Foreign Policy*, Princeton, Princeton University Press.

Krasner, S. D. (1979). 'A statist interpretation of American oil policy towards the Middle East', *Political Science Quarterly*, 94(1), 77–96.

Locatelli, C. (1999). 'The Russian oil industry restructuration: towards the emergence of western type enterprises?' in *Energy Policy*, Issue 27, pp. 435–49.

Logan, J. (2005). 'Energy outlook for China: focus on oil and gas', *Testimony to the US Senate Committee on Energy and Natural Resources*,

Annual Energy Outlook, 3 February, Washington, US Senate, Energy Information Administration.

Lowry, S. T. (1987). *The Archaeology of Economic Ideas*, Durham, Duke University Press.

Majumder, S. (2003). 'The philosophy of privatisation in China', *Business Line*, 3 December, available on-line at http://www.blonnet.com/2003/12/03/stories/2003120300120900.html.

Mannan, M. A. (1989). *Economic Development and Social Peace in Islam*, London, Ta-Ha Publishers.

Marangos, J. (2002). 'A post Keynesian critique of privatization policies in transition economies', *Journal of International Development*, 14, 572–89.

Martin, S. & Parker, D. (1997). *The Impact of Privatisation: Ownership and Corporate Performance in the UK*, London, Routledge.

MBendy (2005). *Africa's Business Watch*, available on-line at http://www.mbendi. co.za/.

Mihaly, P. (1992). 'Property rights and privatisation', *Eastern European Economics*, 31(2), 29–58.

Milani, M. M. (1997). 'Political participation in revolutionary Iran'. In: Eposito, J. L. (ed.) (1997). *Political Islam: Revolution, Radicalism, or Reform?*, Boulder, Lynne Reinner.

Milton-Edwards, B. (2000). *Contemporary Politics in the Middle East*, Cambridge, Politics Press.

Mishra, R. & Navin, B. (2002). *Restructuring Public Enterprises in India: Privatisation and Disinvestment*, New Delhi, Mittal.

Mitchell, J., Seeley, N. & Morita, K. (2001). *The New Economy of Oil, Impacts on Business, Geopolitics and Society*, London, Royal Institute of International Affairs.

Molavi, R. (2003). 'The privatization of the oil and gas industry in Iran', *Economic Trends*, 1(2), 30–33.

Muir (2004). 'Analysis: What now for Iran?'. Report by Jim Muir, BBC Radio Correspondent in Iran, available on-line at http://news.bbc.co.uk/1/hi/world/middle_east/3514551.stm.

Munzen, S. R. (1990). *A Theory of Property*, Cambridge, Cambridge University Press.

Naqvi, S. N. H. (1989). *An Agenda for Islamic Economic Reforms: The Report of the Committee on Islamization Appointed by the Finance Minister, Government of Pakistan*, Islamabad, Pakistan Institute of Development Economics.

Nasri, F. (1983). 'Iranian studies and the Iranian revolution', *World Politics*, 35(4), 607–30.

Nazmi, N. (1989). 'Economic growth and under-development in Iran: 1951–1978', *Comparative Economic Studies*, 31(3), 33–54.

Newbery, D. M. (1992). 'Reform in Hungary: sequencing and privatization'. In: Bird, G. (1992). *Economic Reform in Eastern Europe*, Aldershot, Edward Elgar, pp. 61–77.

Nili, M. (2004). *Industrial Development Strategy of Iran*, Graduate School of Management and Economics, Sharif University of Technology, Tehran, Sharif Technical University Publications.

Nomani, F. & Rahnema, A. (1994). *Islamic Economic Systems*, London, Zed Books.

O'Neill, T. (2004). 'Globalization; fads, fictions and facts', *Business Economics*, 39(1), available at http://www.nabe.com/publib/be/0401/oneill.html.

Ohashi, T. M. (1980). *Privatisation. Theory and Practice*, Vancouver, Fraser Institute.

Owen, R. & Pamuk, S. (1998). *A History of Middle Eastern Economies in the Twentieth Century*, London, I. B. Tauris.

Owen, R. (2000). *State, Power and Politics in the Making of the Modern Middle East*, 2nd edn., London, Routledge.

Pakistani Privatisation Commission (2004). Information on privatisation available on-line at www.privatisation.gov.pk.

Pesaran, M. H. (2000). 'Economic trends and macroeconomic policies in post-revolutionary Iran'. In: Alizadeh, P. (ed.) (2000). *The Economy of Iran: Dilemmas of an Islamic State*, London, I. B. Tauris.

Pesaran, M. H. (2003). Speaking at Royal Institute of International Affairs conference entitled 'Iran', Royal Institute of International Affairs, London, 30 June – 2 July.

PIIC (2000). Petrochemical Industries Investment Company, 'Iran economy from April 1999 to April 2000', available on-line at http://www.irvl.net/iranian_economy_from_april_1999_.html, July.

PIN (2004). Information supplied by the Tehran Information Agency (PIN), available on-line at http://www.petroenergyinfo.net.

Pirie M., (1988) *Privatisation Theory, Practice and Choice*, Aldershot, Wildwood House/Gower.

Poku, N. & Pentiford, L. (1998). *Redefining the Third World*, Basingstoke, St. Martin's Press.

Polanyi, K., Conrad, K., Arenburg, K. & Pearson, H. W. (1957). *Trade and Markets in Early Empires*, New York, Free Press.

Pollack, K. M. (2004). *The Persian Puzzle – The Conflict between Iran and America*, New York, Random House.

Pollack, K. M. (2006). *A Switch in Time: a New Strategy for America in Iraq*, (K. M. Pollack and the Iraq Policy Working Group of the Saban Center for Middle East Policy at the Brookings Institution), Washington DC, Saban Center for Middle East Policy at the Brookings Institution.

Privatisation Organisation (2002). *Laws and Regulations Governing Privatisation in the Third Social, Economic and Cultural Development Plan of the Islamic Republic of Iran*, Tehran, Managing & Planning Organization, Deputy for Administrative, Financial & Human Resources Affairs, Center for Documentation, Museum & Publications.

Privatisation Organisation (2004). *Law of the Fourth Economic, Social and Cultural Plan of the Islamic Republic of Iran 2005–2009 (1384–1388)*, enacted on 1 September 2004, Tehran, Managing & Planning Organization, Deputy for Administrative, Financial & Human Resources Affairs, Center for Documentation, Museum & Publications.

Prizzia, R. (2001). 'Privatization and social responsibility. A critical evaluation of economic performance', *The International Journal of Public Sector Management*, 14(6), 450–64.

Rahnema, A. & Nomani, F. (1990). *The Secular Miracle: Religion, Politics and Economic Policy in Iran*, London, Zed Books.

Rahnema, S. & Behdad, S. (eds) (1995). *Iran after the Revolution: Crisis of an Islamic State*, London, I. B. Tauris.

Ramanadham, V. V. (ed.) (1994). *Privatization and After: Monitoring and Regulation,* London and New York, Routledge.

Reuters (1998a). See www.isreview.org/issues/09/iran_at_crossroads.shtml, footnote 28, 21 April.

Reuters (1998b). See www.isreview.org/issues/09/iran_at_crossroads.shtml, footnote 29, 9 August.

Reuters (1999). See www.isreview.org/issues/09/iran_at_crossroads.shtml, footnote 34, 11 March.

Richards A. & Waterbury, J. (eds) (1996). *A Political Economy of the Middle East*, 2nd edn, Boulder, Westview Press.

Rima, J. H. (1988). 'Keynes' vision and economic analysis'. In: Harmouda, O.F. & Smithin, J.N. (eds) (1988). *Keynes and Public Policy After 50 Years*, Vol. 2, Aldershot, Edward Elgar.

Rima, I. H. (1996). *Development of Economic Analysis*, London, Routledge.

Roshanzamir, N. (2006). 'Iran to privatise but cling to big oil companies', *Financial Times*, 3 June, p. 8. Interviewing Mr Gareth Smith and Ms Najmeh Bozorgmehr in Tehran. Available on-line at http://search.ft.com/searchResults?query/Text=Iran+privatisation+80%25&javascriptEnabled=true&x=12&y=6.

Roy, O. (1994). *The Failure of Political Islam*, London, I. B. Tauris.

Rubin, B. (1981). *Paved With Good Intentions: The American Experience and Iran*, Penguin, Harmondsworth.

Sadr, B. (1989). *The Unitarist Economy*, Tehran, Nasgr-I Hajr.

Saikal, A. (1980). *The Rise and Fall of the Shah*, Princeton, N.J., Princeton University Press.

Samii, B. & Aman, F. (2005). 'Iran: analysts give economic planning and development mixed marks'. Broadcast on Radio Free Europe/Radio Liberty, 26 October.

Shahriari-Rad, P. (2002). *The Privatisation Process in Iran: The Issues of Ownership in Limited Market within an Islamic Tradition and Global Boundaries*, Ph.D. dissertation, Nottingham Trent University, March.

Siddiqi, M. N. (1988). *Muslim Economic Thinking: A Survey of Contemporary Literature*, Jeddah, International Centre for Research in Islamic Economics.

Silverman, D. (2000). *Doing Qualitative Research, A Practical Handbook*, London, Sage.

Smith, A. (1904). *An Inquiry into the Causes of the Wealth of Nations*, Chicago, University of Chicago Press.

Stevens, R. (2004). 'The evolution of privatisation as an electoral policy, c.1970–90', *Contemporary British History*, 18(2), 47–75.

Taleghani, S. (1983). *Islam and Ownership*, Lexington, Mazda.

USEIA (1998). *US Energy Information Administration*, The US Energy Information Administration, U.S. Department of Energy, March. Report available on-line at www.eia.doe.gov.

USEIA (2005). *Country Analysis Briefs: Iran*, The US Energy Information Administration, available on-line at http://www.eia.doe.gov/.

Utilipoint (2004). Utilipoint International Incorporated website, information available on-line at www.energyinfosource.com.

Veljanovski, C. (1989). *Privatisation and Competition. A Market Prospectus*, London, The Institute of Economic Affairs.

Vickers, J. & Yarrow, G. (1989). *Privatization, an Economic Analysis*, Cambridge, Massachusetts, MIT Press.

Vuylsteke, C. (1988). *Techniques of Privatisation of State-owned Enterprises. Volume I: Methods and Implementation*, Washington, DC, World Bank.

Walsh, V. & Gram, H. (1980). *Classical and Neoclassical Theories of General Equilibrium*, New York, Oxford University Press.

Walton, R. E. (1986). *Managing Conflict, Interpersonal Dialogue and Third Party Roles*, 2nd edn, Reading, Massachusetts, Addison-Wesley.

Wells, M. (1999). 'Thermidor in the Islamic Republic of Iran: the rise of Muhammad Khatami', *British Journal of Middle Eastern Studies*, 26(1), 27–39.

Whitehead, C. (1988). *Reshaping the Nationalised Industries*, New Brunswick, Policy Journals.

Wiltshire, K. (1987). *Privatisation: The British Experience*, Cambridge, Blackwell.

World Bank (1995). *Creating Private Enterprises and Efficient Markets*, Development Brief No. 46, January, available on-line at http://www.worldbank.org/html/dec/Publications/Briefs/DB46.html.

World Bank (1997). *Egypt: Issues in Sustaining Economic Growth*, Washington, DC, World Bank.

World Bank (1998). *Country Study. Kyrgyzstan: the Transition to a Market Economy*, Washington, DC, World Bank.

World Bank (2001). *Country Brief*, November, Washington, DC, World Bank.

Wright, R. B. (2000). *The Last Great Revolution: Turmoil and Transformation in Iran*, A. A. Knopf, New York.

Zahlan, R. S. (1998). *The Making of the Modern Gulf States*, Reading, Ithaca Press.

Zonis, M. (1971). *The Political Elite of Iran*, Princeton, New Jersey, Princeton University Press.

Zonis, M. (1983). 'Iran: a theory of revolution from accounts of the revolution', *World Politics*, 35(4), 586–606.

Zubaida, S. (1993). *Islam: The People and the State*, London, I. B. Tauris.

Other Sources

Books and Journals

Amirahmadi, H. & Parvin, M. (eds) (1988). *Post-revolutionary Iran*, Boulder, Colorado, Westview Press.

Anon (2003). 'Reasons behind slow pace of privatization in Iran', *Eqtesad-e Iran*, 5(49), March.

Armajani, Y. (1970). *Middle East Past and Present*, Englewood Cliffs, New Jersey, Prentice-Hall.

Bahgat, G. (2003). *American Oil Diplomacy in the Persian Gulf and the Caspian Sea*, Florida, University Press of Florida.

BBC (2005). 'Iran: an economy on the move?', report by B. Richardson, available on-line at (26/05/2005) http://news.bbc.co.uk/go/pr/fr/-/2/hi/business/4582573.stm.

Beckwith, C.A. & Knox, D. (1984). *Delta Force: the US Counter-Terrorist Unit and the Iranian Hostage Rescue Mission*, London, Arms and Armour Press.

Cerny, P. G. (1979). 'Foreign policy leadership and national integration', *Review of International Studies*, 5(1).

CIA (2005). *World Fact Book: Iran*, Washington, DC, CIA.

Clark, W. (2005). 'Petrodollar warfare: dollars, euros and the upcoming of Iranian oil bourse', *Energy Bulletin*, 9 August.

Economic Trends (2004). 'Propels the private sector', *Economic Trends (Ravand Eqtesadi)*, 2(7), September.

Economist Intelligence Unit (2003). *Iran: Country Profile*, London, The Economist Intelligence Unit.

Economist Intelligence Unit (2004). *Iran: Country Report*, London, The Economist Intelligence Unit.

EIA (2005). *Iran. Country Report*, issued March, available on-line at http://www.eia.doe.gov/emeu/cabs/iran.html.

Emad, E. (2000). 'Privatization and reasons for government's failure to implement it', *Akhbare Eqtesad (Morning Daily)*, 19 February.

Encyclopaedia of the Orient (2004). *Encyclopaedia of the Orient*, available on-line at www.Lexicorient.com.

Handoussa, H. (1997). *Economic Transition in the Middle East: Global Challenges and Adjustment Strategies*, Cairo, Egypt, The American University in Cairo Press.

Henry, C. & Springborg, R. (2001). *Globalization and the Politics of Development in the Middle East*, Cambridge, Cambridge University Press.

Iranchamber (2005). 'Mohammad Khatami', information available on-line at http://www.iranchamber.com/history/mkhatami/mohammad_khatami.php.

Iranian Studies Group (2004). 'Privatization of state-owned enterprises in Iran: panacea or prescription for disaster?', *Iran Analysis Quarterly* (a publication of the Iranian Studies Group at MIT, Cambridge, Massachusetts), September-November.

Khajepour, B. (2005). 'Domestic political reforms and private sector activity in Iran', presented to the Iran Chamber Society, 20 July.

Khatami, M. (1997). *Hope and Challenge: the Iranian President Speaks*, New York, Institute of Global Cultural Studies, Binghamton University.

Khatami, M. (2000). *Islam, Dialogue and Civil Society*, Canberra, Centre for Arab and Islamic Studies (The Middle East and Central Asia), The Australian National University.

Library of Congress (1989). *Iran: a Country Study*, Washington, DC, The Library of Congress.

Milani, M. (1988). *The Making of Iran's Islamic Revolution: From Monarchy to Islamic Republic*, Boulder, Colorado, Westview Press.

Mottahedeh, R. (1986). *The Mantle of the Prophet: Learning and Power in Iran*, London, Chatto & Windus.

Niblock, T. & Murphy, E. (1993). *Economic and Political Liberalisation in the Middle East*, London, British Academic Press.

Parra, F. (2005). *Oil Politics – A Modern History of Petroleum*, London, I.B. Tauris.

Petrov, K. (2006). 'The proposed Iranian oil bourse', *Le Metropole Café*, available on-line at http://lemteropolecafe.com, accessed 16 January.

Ramazani, R.K. (ed.) (1990). *Iran's Revolution: the Search for Consensus*, Washington DC, Indiana University Press in association with the Middle East Institute.

Shafik, N. (ed.) (1998). *Economic Challenges Facing Middle Eastern and North African Countries: Alternative Futures*, London, Macmillan.

Shelley, T. (2005). *Oil-Politics, Poverty and the Planet*, London, Zed Books.

Sick, G. (1985). *All Fall Down: America's Fateful Encounter with Iran*, London, I.B. Tauris.

Stormingmedia (2005). 'U.S. policy regarding Iran in the post-Gulf War era', 5 October, available on-line at http://www.stormingmedia.us/67/6767/A676734.html.

Whitney, M. (2006a). 'Avoiding a war with Iran', available on-line at http://www. opednews.com/articles/opedne_mike_whi_060118_avoiding_a_war_with_.htm, accessed 18 January.

Whitney, M. (2006b). 'Iran's oil exchange threatens the greenback', available on-line at http://www.opednews.com/articles/opedne_mike_whi_060123_iran_92s_oil_exchange_.htm, accessed 18 January.

Whitney, M. (2006c). 'Why Russia betrayed Iran', *Uruknet*, 7 February.

Newspapers and Magazines

Barnameh (weekly magazine), 'Privatization of state-run insurance companies inevitable', 1 February 2003.

Ettela'at (afternoon daily), 'What are the privatization achievements in Iran?', 27 January 2001.

Fars News Agency, 'Stock market may open to foreigners', 23 May 2005.

Fars News Agency, 'Banking privatization in place', 26 June 2005.

Hamshahri (daily newspaper), (interview) 'Not seek profiteering by privatization', 10 June 2002.

ILNA News Agency, 'Insurance firms favor privatization', 21 September 2004.

ILNA News Agency, '421m shares on offer by March 17', 11 March 2005.

ILNA News Agency, 'Private sector will get $8b', 16 March 2005.

ILNA News Agency, 'Rail privatization scheme ready by September', 15 June 2005.

ILNA News Agency, 'Iran air privatization on agenda', 24 June 2005.

Iran Daily, 'Privatization at snail's pace', 25 August 2004.

Iran Daily, 'Privatization bill in Majlis by March', 28 August 2004.

Iran Daily 'No alternative to privatization of banks', 31 August 2004.

Iran Daily, 'Ownership rights and wealth generation', 6 September 2004.

Iran Daily, 'SEC still studying key constitutional amendments', 14 September 2004.

Iran Daily, 'Economic reforms move backward', 16 September 2004.

Iran Daily, 'Khatami sets privatization deadline', 22 September 2004.

Iran Daily, 'Privatization organization not repaying state debts', 6 October 2004.

Iran Daily, '12 major state firms on tender', 12 October 2004.

Iran Daily, 'Private sector getting fair share of Forex Fund', 13 October 2004.

Iran Daily, 'TSE offers shares of 4 major state firms', 18 October 2004.

Iran Daily, 'SEC decision to boost privatization', 23 October 2004.

Iran Daily, 'Privatization setbacks due to resistance', 24 October 2004.

Iran Daily, 'Privatization organization defends performance', 27 October 2004.

Iran Daily, 'Government advised to abandon business approach to privatization', 30 October 2004.

Iran Daily, 'Privatization organization urged to review share values', 1 November 2004.

Iran Daily, '14 more state firms face privatization', 6 November 2004.

Iran Daily, 'Privatization chief claims success', 20 November 2004.

Iran Daily, 'Privatization ineffective without improved productivity', 21 November 2004.

Iran Daily, 'SEC okays steel privatization initiative', 24 November 2004.

Iran Daily, 'Rls.15 trillion from privatization next year', 26 November 2004.

Iran Daily, 'Privatization underscored', 29 November 2004.

Iran Daily, 'Privatization of petrochemical complex illogical: sector needs financial resources', 18 December 2004.

Iran Daily, 'Rls.400b for privatization', 24 December 2004.

Iran Daily, 'Where privatization stands', 30 December 2004.

Iran Daily, 'Privatization lacks transparency', 11 January 2005.

Iran Daily, 'Economy needs private sector to survive', 15 January 2005.

Iran Daily, 'Low-profile privatization only part of the problem', 31 January 2005.

Iran Daily, 'Privatization defied!' by Farzaneh Rahimzadeh, 5 February 2005.

Iran Daily, 'Privatization goals lagging', 21 February 2005.

Iran Daily, '25% of privatization revenue target realized', 26 February 2005.

Iran Daily, 'Renewed call for liberalization', 26 February 2005.

Iran Daily, 'CBI privatization opposed', 11 April 2005.

Iran Daily, 'Red-lines hurt private investment', 1 May 2005.

Iran Daily, 'Moving along privatization', 3 May 2005.
Iran Daily, 'Privatization process expects major boost', 6 May 2005.

IRAN Newspaper, '42m shares ceded to private sector', 20 September 2004.
IRAN Newspaper, 'Growth prospects improve', 25 September 2004.
IRAN Newspaper, 'Urban development & private sector', 16 May 2005.
IRAN Newspaper, 'Stock market spurs economic growth', 29 May 2005.
IRAN Newspaper, (interview) 'Inadequate planning', 1 June 2005.
IRAN Newspaper, 'Private sector gets $6b for industrial projects', 15 June 2005.
IRAN Newspaper, 'Privatization & constitution', 19 June 2005.
IRAN Newspaper, 'Private banks', 25 June 2005.

IRNA News Agency, 'Government ready to privatize several banks', 16 November 2004.
IRNA News Agency, 'State-owned insurance companies to be privatized next year', 22 December 2004.
IRNA News Agency, 'Reducing banking rates can boost economic vibrancy', 2 January 2005.
IRNA News Agency, 'Khatami outlines details of draft budget bill for 1384', 9 January 2005.
IRNA News Agency, 'State-owned companies shares sold has increased 740 percent', 26 January 2005.
IRNA News Agency, 'Privatization revenues not to be included in next year's budget', 6 February 2005
IRNA News Agency, 'US threats not a hurdle to foreign investments in Iran', 26 February 2005.
IRNA News Agency, 'Iran privatization performance', 24 April 2005.
IRNA News Agency, 'Development not possible without privatization', 28 April 2005.
IRNA News Agency, 'Iran economic development hinges on prudent investment', 19 May 2005.

ISNA News Agency, 'Privatization pros and cons', 22 September 2004.
ISNA News Agency, 'Private insurance firms okayed', 24 September 2004.
ISNA News Agency, '450m shares sold in 12 months', 4 March 2005.
ISNA News Agency, 'Privatization organization committed to share offer', 8 March 2005.

ISNA News Agency, 'Banks prepare for privatization', 29 May 2005.
ISNA News Agency, 'Anti-monopoly bill ready for parliament', 6 June 2005.
ISNA News Agency, 'State firms sell 124m shares in 3 months', 10 June 2005.

Moj News Agency, 'Boosting public participation', 22 May 2005.

Sobhe Eghtesad (Newspaper), 'Privatization in Iran: ahead challenges and barriers', 11 April 2004.

INDEX